ESSENTIAL
METALLURGY
for
ENGINEERS

W.O. Alexander, PhD, CEng, FIM, is Professor Emeritus in the Department of Metallurgy, University of Aston

G.J. Davies, BE, MA, PhD, ScD, CEng, FIM, MIMechE, is Professor of Metallurgy at the University of Sheffield

S. Heslop, BSc, CEng, FIM, is a Consultant

K.A. Reynolds, MSc, ACT, CEng, FIM, MIMM, is a Consultant

V.N. Whittaker, MSc, CEng, MIM, is a Visiting Lecturer at Lancaster University, Engineering Department

E.J. Bradbury, MEng, CEng, FIM, MIMechE, is a Director at BNF Metals Technology Centre, Wantage

ESSENTIAL METALLURGY for ENGINEERS

W.O. Alexander, G.J. Davies, S. Heslop
K.A. Reynolds, V.N. Whittaker

Edited by
E.J. Bradbury

Van Nostrand Reinhold (UK)

© 1985 W.O. Alexander, G.J.
Davies, K.A. Reynolds

First published in 1985 by
Van Nostrand Reinhold (UK) Co. Ltd
Molly Millars Lane, Wokingham,
Berkshire, England

Typeset in 10/11 pt Sabon by
Columns Ltd, Reading

Printed in Great Britain
by The Thetford Press Ltd,
Thetford, Norfolk

Library of Congress Cataloging in Publication Data
Main entry under title:

Essential metallurgy for engineers.

Includes index.
1. Physical metallurgy. 2. Engineering design.
I. Alexander, William Oliver. II. Bradbury, E.J.
TN690.E85   1985   620.1'6   84–25600
ISBN 0-442-30624-5

Contents

WITHDRAWN
UTSA LIBRARIES

Preface

The whole philosophy of engineering design is contained in Holmes' *The Deacon's Masterpiece* written in 1775:

> There is always somewhere a weakest spot . . .
> And that's the reason, beyond a doubt,
> A chaise breaks down, but doesn't wear out . . .
> "Fur,", sed the Deacon, 't's mighty plain
> Thut the weaks' place mus' stan' the strain;
> "N" the way "t" fix it, "uz" I maintain,
> Is only "jest",
> "T" make that place "uz" strong "us" the rest.

In the Deacon's time design was a question of trial and error, and failure prevented by increasing the appropriate dimensions at the points where failure occurred.

Today trial and error is no longer a practical method of design and the Deacon's objective is achieved by the application of proven design theory and the use of construction materials selected from the wide range available. Failure may and does still occur and is now prevented from recurring by again changing dimensions, but also possibly changing the material of construction or the method of manufacture.

Both design theory and materials available have developed and become increasingly complex and it is no longer possible for a single individual to be competent in both aspects of engineering. It is, however, very necessary that the engineers responsible for the design and the engineers responsible for the material can communicate and have an understanding of each other's work.

It is the purpose of this book to introduce design engineers to an understanding of metallic materials which will enable them in turn to understand the work of the materials engineer and what that engineer can do to assist them.

E.J.B.

Acknowledgements

The authors are grateful to the following for permission to reproduce the illustrations indicated:

American Society for Metals, *Metals Handbook* Vol. 9, 8th edn: Figures 3.14, 3.30, 4.5, 4.7(a), 4.9, 4.16, 4.23; *Source Book in Failure Analysis* (1974): 4.4, 4.7(b), 4.11, 4.17(b), 4.20(a), 4.21

Edward Arnold (Publishers) Ltd, *The Plastic Deformation of Metals*, R.W.K. Honeycombe (1968): Figure 4.8(b)

British Steel Corporation: Figures 5.1, 5.2, 5.3, 5.4

Carl Hanser Verlag, West Germany, *An Atlas of Metal Damage*, L. Engel and H. Klingele (1981): Figures 4.6, 4.8(a), 4.12, 4.17(a), 4.18, 4.20(b), 4.22, 4.24

Dr. Riederer-Verlag, West Germany: Figure 4.7(b)

Gordon & Breach, New York, *Why Metals Fail*, R.D. Barer and B.F. Peters (1970): Figure 4.15

Institute of Metallurgists: Figure 2.12

Magnaflux Ltd: Figure 6.25

Metals Society, *The Microstructure of Metals*, J. Nutting and R.G. Baker (1965): Figures 3.33, 3.34; *Non-metallic Inclusions in Steel*, R. Kiessling and N. Lange (1978): Figure 3.23

Nato Science Committee Study Group: Figure 7.6

Van Nostrand Reinhold (UK), *An Introduction to the Properties of Engineering Materials* 3rd edn, K.J. Pascoe (1978): Figure 3.31

Wells Krautkramer Ltd: Figures 6.18, 6.21

John Wiley, Inc., New York, *Analysis of Metallurgical Failures*, V.J. Colangelo and F.A. Heiser (1974): Figures 4.14, 4.19

Introduction

Even as late as the 19th century, construction materials used in engineering applications were simple — wood, leather, cast or wrought iron, brass, bronze — and production methods were just as simple and traditional. Design was largely a question of trial and error and experience; failure was regarded as merely unfortunate even if it resulted in casualties, and at least it provided some lessons to be learned. Legislation was almost non-existent, and mass production in the modern sense a thing of the future.

Since the turn of this century the situation has changed so markedly that what went before is of interest only to historians rather than providing useful experience for engineering designers today. The pace of progress in science and technology, the need to conserve resources, the pressures of commercial competition, the burden of increasing legislation and the complexity of modern materials all contribute to the fact that, today, no individual working in isolation can hope to design a fully satisfactory component. Design is now a matter of close cooperation between several individuals, each contributing his own expertise and experience. In particular, it is the result of cooperation between the design engineer and the materials engineer. The design engineer is responsible for the geometry of the design; the materials engineer for the selection of the materials and manufacturing routes.

The final design is inevitably a compromise between cost and the likelihood of failure (and cost must include an assessment of the consequences of failure). The integrity of any structure or machine is the integrity of its weakest component — the one that fails first. Failures may be caused by an inability to meet a competitor's price, errors in dimensioning, inadequate assessment of a corrosive environment, lack of knowledge of the effects of low or elevated temperatures, or by use of an inappropriate manufacturing technique. The consequences of failure can be so minor that they have only a nuisance effect, or so great that lives are endangered or the continued existence of a company threatened. There are many instances of annoying failures — handles that come off, locks that jam — but the effects of major failures are not always so well known; it is recorded that the cost of a failure of a coffee percolator handle was $35 million in recall and replacement charges!

We have said that modern materials — particularly metals — are becoming more and more complex, but why is this? In earlier days

machinery moved at walking pace. The metals available were of comparatively low strength and the dimensions of components were large and
problems of stiffness (that is, rigidity) did not arise. With the advent of very
high speeds of rotation and linear movements in modern equipment, the
increase in the forces to be withstood or transmitted by a component, and
the increase in the frequency of application of those forces, have required
the development of metals of high strength, and high fatigue- and impact-
resistance. The consequence of the economic pressure to use high-strength
but costly materials to the limit of their capacity forced design engineers to
reduce cross-sections of components and this, in turn, introduced further
design problems associated with maintaining stiffness in the component,
and increasing difficulties with metal fatigue and corrosion.

High-strength metals are now far more complex than their weaker
predecessors and for their economic use require a thorough understanding
of their strengths and limitations. Thus, high-strength steels are rarely
weldable and can be less corrosion-resistant than lower strength steels.
High strength is not the only reason for complexity, however. The need for
metals that resist deformation at high temperatures has introduced a now
extensive range of complex creep-resistant alloys, and the need to retain
strength at high temperatures in corrosive environments has led to even
further complexities. The range of alloys available to the designer has
increased through the use of metallic elements other than the copper and
iron used in the cast and wrought iron, brass and bronze available in
earlier times. Thus aluminium, magnesium, nickel, chromium and other
metallic elements are now in common use in the very wide range of
materials to be considered by the design engineer today.

Paralleling this wide range of materials is an equally wide range of
possible manufacturing routes. The processes available for the manufacture of a component and which will affect its final design differ widely in
character, in the metals with which they can be used, and in the effects
they have on the properties of the finished component. Such processes
include casting to shape, machining from the solid, welded fabrication,
pressing to shape, hot or cold forging and powder metallurgy. Each
manufacturing process brings with it both advantages and disadvantages
and each has a radical effect on the material being processed and the kind
of subsequent operations that need to be performed.

With any design the first thing to do is to define the complete function of
the component. The word 'complete' means not only the magnitude of the
forces it must withstand and transmit, but also what physical and chemical
characteristics it must have. The *criticality* of the component must also be
assessed — that is, what is the cost of failure? The commercial aspects
must be known along with any restrictions on production processes that
may be used. Thus, the availability of foundry facilities in a company's
works may, at an early stage in the design exercise, suggest that
consideration of a powder metallurgy production route for a component
would not be practical.

All these factors must be considered, preliminary decisions arrived at,
reconsidered in the light of changes which prove to be essential in

geometry, processing, properties and so on, and the cycle repeated until prototypes are produced. Simultaneously, questions of inspection, quality assurance and testing must be examined and decisions made. Eventually, a review of actual service experience may well require a re-examination of the whole design sequence, for the ultimate test of any component is its performance under service conditions.

Clearly, it is vital for the design engineer to have an appreciation and understanding of the problems facing the materials engineer in his selection of the material and processing route for a component. The continuous dialogue between the design and materials engineers is illustrated in Fig. 0.1. Notice that design is a cyclic process which does not necessarily end when a component with satisfactory performance is produced.

The complexity of the problem in selecting a metal to be used for the manufacture of a component becomes apparent when considering the very broad range of the properties of metals and the 'materials' loop of the design process. The engineer concerned in the 'design' loop must be aware of this complexity and understand the nature if not the detail of the problems being faced.

Below is a list of important, broad, characteristic properties of metals (to which can be added cost and availability):

mechanincal strength
malleability and ductility
hardness and wear resistance
toughness
corrosion resistance
electrical properties
magnetic properties
colour
joinability
ease of working

Each of the above properties is itself a function of contributory properties. Thus, considering only the mechanical performance of a metal, this is a function of its

elastic moduli
tensile, compressive or torsional properties
fracture toughness
creep resistance
fatigue resistance
strength at high or low temperatures

Further, within each of these categories there are further variables such as yield point, Poissons ratio, notch sensitivty and so on.

The number of a metal's properties to be examined is therefore considerable and each property varies, both in itself and in its relation to

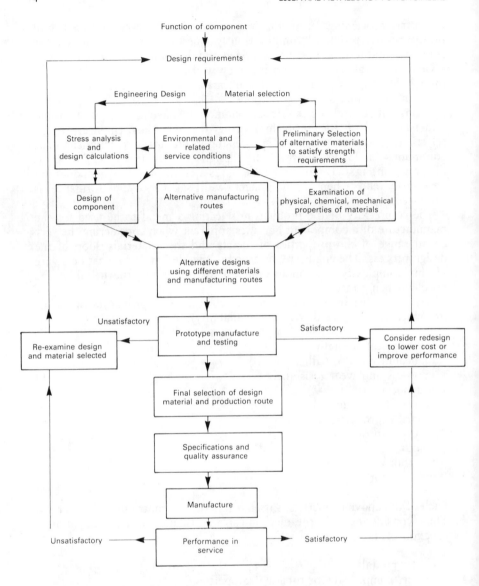

Fig. 0.1 The iterative nature of the design loop showing interactions involving materials selection and performance

other properties, according to changes in composition, in manufacturing process, or in both.

For each variation in a particular property of a metal being used in the design of an item there will be a corresponding variation in the geometry of the final product, emphasizing the importance of a clear and accurate specification of the design requirements. Similarly, a change in the shape or dimension of the final product may well require a change in an essential property of the material from which it is to be constructed.

With a wide and complex range of metals and properties, design and materials engineers are fortunate that by following a comparatively simple procedure the number of alternatives available can be drastically reduced. The procedure is to match each material against four basic characteristics defined in the design specification:

geometry
stress
environment
processing

It is to be noted that all four characteristics are of concern in the selection of a metal.

A successful product will result only from close understanding and cooperation between the design and the materials engineer and for these to be achieved it is essential that each should know the basis on which a decision by his partner is made.

It is the intention in the succeeding chapters to provide the design engineer with an understanding of the origins of the strength and weaknesses of metals, and descriptions of the properties that are of importance in a design exercise, how these can be examined to ensure the quality of a product, the performance of metals in service, and economic factors concerned in the selection of a metal and a processing route. With this understanding cooperation with the materials engineer will be readily achieved and a successful product result.

CHAPTER I

Materials selection, manufacture and engineering design

There are many ways in which it would be possible to generalize about designing an engineering component but the particular purpose here is to show the stages at which material selection is likely to be considered. Figure 0.1 (page 4) is structured to show these stages and the cyclic reiterative nature of the design, materials and processing selection procedures. The object of the whole design exercise is to satisfy some desired function within a stated set of constraints, whether the need be for a paper clip or a space satellite. Even the most complex design can be broken down into components, each of which can be assigned specific functional requirements and thus start off its own design loop. However, the interaction with other components in the design will repeatedly intrude into that loop. For example, a component that has to be joined to another by welding cannot be designed or materials chosen without regard to their interactions. Figure 0.1 is, therefore, an idealized representation of the design loop for a single product, and its main purpose is to show how the material chosen and the manufacturing route are inseparable from all of the other technical functions and constraints.

In practice, not surprisingly, several of the internal loops shown in Fig 0.1 may be short-circuited.

1.1 SPECIFICATION OF DESIGN REQUIREMENTS

To arrive at a satisfactory end-product it is clear from Fig. 0.1 that the first essential is a detailed and complete specification of design requirements. Four basic questions need to be asked, and answered:

is there a market?
is the design suitable?
is the manufacturing process right?
is the quality acceptable?

It is the last three questions that are of particular concern here but as all are interrelated it is necessary to consider all four.

1.1.1 The market

Without a market there can be no sale; for a market to exist certain conditions must prevail:

> there must be a demand for the end-product;
> the product must satisfy all the end-use requirements;
> the price must be acceptable;
> the quality must be acceptable;
> the delivery time must be acceptable.

Let us consider each of these conditions.

Demand

Where a demand for a product already exists, the market for a new design will depend on its satisfying the four remaining conditions. In the case of a new design the conditions that will create a market are outside the scope of this book.

End-use requirements

A product that does not satisfy all the end-use requirements will have no market or, if the deficiency becomes apparent after some time in service, the market will disappear. There are many examples of how a series of mechanical failures of components, such as the rusting of bright decorative items, has led to a serious loss of sales which could have been avoided if closer attention had been given to design, manufacture or quality.

The technical specification for a component will frequently not define all the requirements it must satisfy. In many cases the requirements are not readily quantified, yet have a significant effect on sales and performance. For example, the technical specification for a car bumper may define dimensions, form, material, finish, performance against impact, abrasion, corrosion and so on. However, in addition, the customers who buy the car will expect it to:

> be aesthetically pleasing,
> require no maintenance,
> be easy and cheap to replace,
> maintain its appearance for the life of the car.

Failure to meet these unquantifiable requirements will lead to advertised dissatisfaction of the customer, affecting sales, with consequent low part-exchange value again affecting sales.

Strength and mechanical property requirements can in almost all cases be specified numerically, but other technical requirements are not so readily defined. Thus the materials used for turbine blading in a land-based gas turbine may well be subjected to the same stresses and temperatures as those used in a similar application in an aircraft gas turbine, but the fuel used in the land-based unit is almost certainly less refined than that used in the aircraft unit and will bring with it the danger of serious corrosion. Such corrosive conditions cannot be defined numerically but, being specified as 'existing' must be considered and will almost certainly lead to the selection of different blading materials for the two applications despite the fact that the strength requirements are practically the same.

It is essential, therefore, that the specification for a component not only defines the mechanical requirements that must be satisfied, but also defines all aspects of the application the component must satisfy.

Price

Little comment is required except to emphasize that as material and manufacturing costs are important factors in the calculation of selling price, these two costs must be considered carefully during the design of a component.

Quality

Quality is intimately connected with price and performance. Some components or assemblies can be regarded as 'throw-away' items whose life-span is not important but whose cheapness is. Thus, quality requirements must be determined before design or material selection commences. A safety match manufactured in teak and machined to close tolerances is clearly not a marketable item!

Delivery time

The questions to be answered are: is it possible to design, develop, and manufacture the particular item for sale during the period when a market for it exists?; are competing products likely to steal the market? (this is particularly essential with highly fashionable goods).

These five conditions that primarily control the market for any product must be taken into consideration before design commences and their influence on the design assessed.

1.1.2 Legal liability and consumer protection

Another increasingly important market specification of a technical nature is the need to satisfy consumer protection and product liability legislation. In the past the buyer or consumer had to prove the product was of 'unmerchantable quality' or that a manufacturer or agent had been 'negligent' in some way that resulted in damage, loss of use or injury to the consumer. Nowadays the onus is on the manufacturer, and his design team have to prove to the Courts that they took all reasonable steps to produce a satisfactory and a safe product. It is not enough to design only for the obvious use, but *all* foreseeable misuses must also be taken into account and records kept to establish that this was done. To take a simple case, a manufacturer of screwdrivers must not only guard against failure when the implements are used for the intended purpose of fixing or undoing screws, but must also foresee their use for opening tins of paint, punching holes and so on.

Failure to take account of such issues in the design stage can result in legal damages being awarded at a level that could cripple a company. It is surprising how ignorant technical personnel appear to be of the implications of omitting to take proper account of these legal obligations.

Consumer legislation has already made a significant impact on design and materials selection in fields such as automobiles and domestic products. Examples are the simple precautions taken to eliminate sharp or hard shapes likely to injure pedestrians in accidents, and the replacement of lead domestic water piping by copper or plastic piping. A further example is the incorporation into the design of cars of compartments that collapse progressively under collision impact so as to absorb energy that could otherwise injure passengers in the vehicle. These changes all reflect the need to satisfy changing market requirements.

1.1.3 The design specification

The specification defines the product in terms of the geometry of the component parts, the way they are arranged, their dimensions, required properties and the materials of construction. Its purpose is to ensure that the product performs its intended function and can be sold at an economic price. The latter is particularly important for mass-produced articles for the consumer market, but less so in small quantity manufacture or items for military purposes. The selection of appropriate materials is most critical where the application is cost sensitive.

When the functional and mechanical aspects for a particular item have been defined, the selection of a suitable material depends upon several factors, including chemical, physical and mechanical properties, suitability for the proposed manufacturing method, stability in service over the anticipated lifetime and, related to all of these, cost and availability. Like the design process itself, selection of the material is an interative process that takes account of past experience and demands judgement in optimizing the combination of properties in the material finally chosen. The way in which strength and mechanical property data are used will be dealt with in Chapter 3.

Certain applications involve less demanding criteria on choice of material than others. For example, the housing of a lighting cluster on a motor car is basically something that requires little in the way of mechanical strength, but must have an attractive appearance and retain this for the life of the vehicle; its purpose is as much decorative as functional. Such an article could be designed in various ways from totally different materials and manufacturing methods:

(i) Die casting in aluminium, or zinc alloy, electroplated to impart corrosion resistance and aesthetic appearance.

(ii) Press formed from brass or steel sheet, polished, electroplated or plastic coated.

  (iii) Injection moulded, plastic, self coloured, aluminized, or plated.
  (iv) Combinations of materials similar to all three of the above, including the moulding-in of conductors into plastic mouldings.

Whichever material and process is chosen has implications on the design. For instance, if metal, then there will be no problem with the earthing electrical contact, but the 'live' will have to be insulated; if plastic, provision will have to be included to make both earthing and live contacts to the electrical supply of the vehicle. Again, the shape finally chosen must reflect the forming characteristics of the material selected and take into account whether thread forms can be machined *in situ* or moulded into the shape as inserts, or some other method of fastening used, such as integral clips. Machining would be avoided to minimize costs and self colouring or aluminizing would be cheaper than electroplating or painting. For such a design, the most significant criterion would be the final cost in relation to aesthetic and functional requirements, and the choice of material would be dominated by the overall production cost.

    In contrast to the above example the selection of material for a load-bearing component would take much more account of the mechanical properties such as strength and toughness, and the manufacturing method would be largely determined by the shape and properties demanded. For example, for the pistons for a high-performance engine there is virtually only one type of material — a high-strength lightweight aluminium alloy having good casting and machining properties, to allow the complex shape to be produced, together with good corrosion resistance and ability to maintain its strength at the working temperature of the engine. Materials for applications such as high-tensile wire rope, tanks and pressure vessels subjected to corrosive liquids, electric conductors and electronic devices, and components in nuclear reactors in effect select themselves by virtue of their unique chemical, physical or mechanical properties. In such situations the manufacturing method and final costing become of secondary importance and the design specification is dominated by the need to fulfil the technical function.

    In the last few decades many attempts have been made to devise systems for selecting an appropriate material, but with limited success. The problem is that each design presents its own unique combination of requirements and priorities with the result that, although it is a fairly straightforward matter to arrive at a short list of a few candidate materials, the final selection depends on subtleties of the individual design and the current 'state of the art' in manufacture, as well as cost and availability. These aspects will be further elaborated upon in later chapters, but it is worthwhile indicating some of the pitfalls of generalized data on materials.

    In the 1960s, Alexander pioneered the concept that designers were fundamentally buying 'properties' when they sought a material for a particular purpose, and because the traditional approach was to consider materials within narrow categories like 'metals', 'plastics', 'timber', 'concrete', and so on, it was necessary to adopt a much broader approach

Table 1.1 Cost per unit of property (after Alexander, 1979)

Material	Tensile strength (MN m^{-2})	Modulus of elasticity (MN m^{-2})	Fatigue strength (MN m^{-2}) 10^8 cycles	Density (kg m^{-3})	Cost (£ tonne^{-1})	Cost per unit of MN m^{-2} (£)		
						Tensile strength	Modulus of rigidity	Fatigue strength
Cast iron (castings)	400	45 000	150	7300	500	9.1	0.08	24.3
Steels								
mild steel–free cutting (bar)	360	77 000	193	7850	244	5.3	0.020	9.9
high tensile steel 1.5Ni–Cr–0.25Mo (bar)	1000	77 000	495	7830	400	3.1	0.04	6.3
austenitic stainless 304 18Cu–8Ni (sheet)	510	86 000	250	7900	1 500	23.2	0.140	4.7
Non-ferrous metals								
brass 60Cu–40Zn (bar)	400	37 300	140	8360	1 000	20.9	0.230	59.7
aluminium alloy (sheet)	300	26 000	90	2700	1 000	9.0	0.100	38
duralumin (sheet)	500	26 000	180	2700	2 000	11	0.210	30
magnesium alloy (bar)	190	17 500	95	1700	3 700	32	0.350	66
titanium 6Al–4V (bar)	960	45 000	310	4420	11 000	50	1.100	157
Plastics								
polyethylene	13	84	3.25	920	535	38	—	151
Nylon	86	2 850	20	1140	1 705	23	0.680	97
PVC (rigid)	50	1 680	12.5	1400	445	12.5	0.370	50
Reinforced concrete beam	38	10 000	23	2400	20	1.26	0.005	2.1
Timber								
softwood	5*	2 000	3	550	250	27	0.070	46
hardwood	14*	4 500	6	720	400	19	0.060	45

*Simple tensile force along grain

to selection that transcended traditional barriers between competing materials. Accordingly a 'cost per unit of property' was proposed as a way of making a comparison, and thence a selection. Table 1.1 indicates the basis of the approach. All of the costs underlying such data need to be continually reviewed in the light of changes in price, particularly of the raw materials and semi-manufactured forms since these reflect market conditions and can alter markedly over a short period. However, the major limitation to such approach is not connected with the economic aspects (in fact the Table shows how easy it is to take account of such changes despite the widely differing materials and sources), but with optimizing the *combination* of properties required for a particular design. Concrete and grey cast iron are both castable to complex forms and are both brittle in tension and strong in compression, but what designer would contemplate making the cylinder block of an engine in concrete? Or a connecting rod or crankshaft for an engine from plastic? Or timber for the frame of a high-temperature furnace? These examples highlight the dilemmas characteristic of practically every attempt to systematize material selection, though all are valuable in focusing attention on a short list of possible choices and some are particularly suited to particular fields of design.

Table 1.2 Strength and stiffness of engineering materials

Material	Tensile strength (C) (MNm^{-2})	Youngs modulus (E) (GNm^{-2})	Density (ρ) (kgm^{-3})	Specific strength (C/ρ) ($\times 10^4$ NMkg^{-1})	Specific stiffness (E/ρ) ($\times 10^7$ Nmkg^{-1})
Cast iron	200	110	7150	2.80	1.54
Steel					
mild	450	210	7860	5.72	2.67
strong alloy	1500	210	7800	19.23	2.69
stainless	500	210	7930	6.31	2.65
Aluminium					
pure	70	70	2710	2.58	2.58
strong alloy	450	70	2800	16.01	2.50
Copper (pure)	140	120	8930	1.56	1.34
Brass (70 Cu–30 Zn)	400	120	8500	4.71	1.41
Magnesium	250	42	1740	14.36	2.41
Titanium alloy	1200	120	4580	26.20	2.62
Plastics					
polyethylene	13	0.18	925	1.41	0.02
melamine	75	9.0	1500	5.0	0.60
rigid PVC	60	2.8	1650	3.6	0.17
Timber					
softwood	60*	12	500	12.0	2.40
hardwood	120*	15	750	16.0	2.00
Concrete	5	14	2400	0.21	0.58
Glass-fibre					
Reinforced plastic	500	60	2200	22.72	2.72

*Maximum tensile force in outer fibres during bending

Technical criteria or '**merit indices**', such as specific strength ($\frac{\text{yield strength}}{\text{density}}$) and specific stiffness ($\frac{\text{Young's modulus}}{\text{density}}$), are useful if a material is being considered for load-bearing applications where weight must be minimized. Table 1.2 gives some comparative values, which show that widely differing materials, such as timber and steel, are in fact quite similar in terms of specific strength and stiffness. It is also rather surprising to see how close are the values for the common high-strength engineering metals, iron, aluminium, magnesium and titanium.

Although data like those in Table 1.2 are undoubtedly useful in focusing attention on candidate materials, the final choice is usually determined by properties other than the parameters given. In the final selection it may be some quite different property, such as corrosion resistance, creep resistance or compatibility with the service environment that overrides the specific properties. Some factor, such as availability in a desired form, weldability or working properties, may preclude the materials that are top in the strength ranking and, as in the 'cost-per-cent of property' approach, it may be that the combination of properties or the need for some special requirement, such as toughness or electrical conductivity, will dominate the selection.

Finally, when tables of properties are studied, it must be borne in mind that they usually refer to ambient or some specific temperature, and properties ranking of all materials may alter drastically at different temperatures. The ductile–brittle transition temperature of low carbon steels is not far below room temperature and many failures have occurred when steel components containing small notches, cracks or re-entrant geometries, that had more than adequate toughness at temperatures in the region of 20°C were, perhaps only on an isolated occasion, subjected to shock loads at mildly sub-zero temperatures. At the other end of the scale, high-strength aluminium alloys may weaken dramatically when exposed for moderate periods at temperatures in the region of 200°C, whereas a steel to which they may appear to be superior on a strength-to-weight basis is hardly affected by the same exposure and can be safely used at temperatures of over 400°C without any loss of strength.

In very broad terms, the influence of temperature on the strength of a metal can be gauged from the *homologous temperature*, which is the ratio of the service temperature to the melting temperature (T_m) in kelvins. Figure 1.1. shows how strength falls only slowly over the range up to 0.4 T_m, but drops steeply above 0.6 T_m. In the region 0.4–0.6 T_m, the properties tend to be time dependent and this is where creep resistance has to be considered. In the previous comparison of aluminium ($T_m = 930K$) and steel ($T_m = 1800K$), it will be realized that 20°C represents 0.31 T_m for aluminium and only 0.16 T_m for steel, whereas 200°C represents 0.5 T_m and 0.26 T_m, respectively. To an Eskimo, ice blocks appear to be an attractive material for construction of a house, but not to those of us who live in temperate climates. So we should be equally cautious when comparing room temperature properties of materials with a view to selecting for a type of service environment that may involve changes in

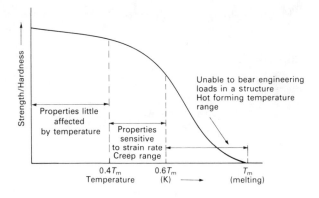

Metal	T_m (K)	Melting temperature (C)	Approximate limiting service temperature (C)
Tin	505	232	30
Lead	600	327	90
Zinc	692	419	140
Magnesium	923	650	280
Aluminium	934	661	290
Copper	1356	1083	540
Nickel	1728	1455	760
Iron	1810	1537	820
Titanium	1938	1665	880
Tungsten	3685	3410	2000

Fig. 1.1 Effect of melting temperature on strength

temperature. The homologous temperature assessment is a very useful criterion, and an indication of the useful temperature range for common engineering materials is given in Fig. 1.1 in the last column.

Although T_m strictly applies only to metals, and gives only a rough guide with alloys developed for specifically enhanced properties, such as high-temperature strength or creep resistance, the concept of comparing the service temperature with the melting point gives a useful indication of whether or not attractive properties can be usefully exploited. Even those substances that have no well defined melting point are hardly likely to be able to support loads at temperatures where they sublime, ignite or undergo a glass transition. Moreover, although metals are known to corrode or oxidize in certain environments, the non-metals too are subject to their own peculiar forms of degradation in service, from biological attack of timber to ultraviolet degradation and environmental stress cracking of plastics. No material is immune from some type of deterioration in some possible service environment.

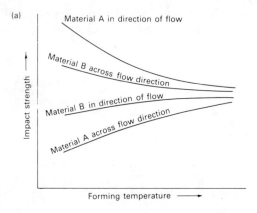

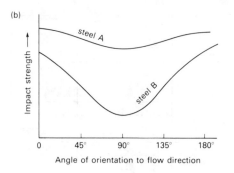

Fig. 1.2 Variation in impact strength in relation to fibre. (a) High impact polystyrene mouldings; (b) two different qualities of low carbon steel

1.1.4 The manufacturing specification

Although their aims are quite distinct, the specifications for material, manufacturing process and quality are in practice closely interwoven and it is impossible to consider any one in isolation from the other two. The purpose of the manufacturing specification is to set down all the essential requirements for making an article, beginning with the raw material or stock (which in turn has its own set of specifications), and including all the required details of form, dimensions, tolerances, surface finish, forming processes and heat treatment; in fact, everything necessary to enable the product to be manufactured to fulfil its design requirement.

The interplay between the three specifications can be well illustrated by considering a simple shape that is to be cut from steel plate to some kind of profile. If the calculations made by the designer are based on the properties of low carbon steel, then the profile will almost certainly be flame cut from the plate. However, the selection of the cutting process is dependent upon the choice of material; if a medium or high carbon steel had been specified, then flame cutting would leave an unacceptable hard metal edge, and if a non-ferrous material had been specified then flame cutting would be unsuitable and the plate would have to be machined. Now the properties of low carbon steel plate are known to be highly directional, arising from the fact that the plate has been made from a large piece of cast metal that has been subsequently processed into a smaller cross-section. In this process, non-metallic inclusions and compositional variations within the steel give rise to a grain or fibre structure (which will be more fully explained in Chapter 2).

The properties associated with the material in practically all the commercial specifications are, however, those associated with the longitudinal direction, that is properties measured on test pieces with the grain running in an axial direction and generally representing the best combination of strength, ductility and toughness. The quality of the steel plate has a large influence on the difference between the longitudinal properties and the transverse. In 'poor' quality steels, that is with many inclusions that affect ductility and toughness and have an unfavourable grain structure, the transverse properties may well be less than half those of the longitudinal. This fact is well recognized by producers and users of such steel, but very seldom appears in the material specification. Nevertheless, when the shape is cut from the steel plate its functional properties will depend very greatly on how the template for the profile is placed on the steel before cutting. If care is taken to see that the grain structure of the plate runs parallel to the critical axis of the profile, then the optimum properties may be realized in the shape, but if it is not so suitably oriented, then the cut shape may have toughness or fatigue resistance far below what the designer expected and intended. Hence the manufacturing specification in this instance not only has to be suited to the material chosen for the application, but must also set down precise details of how the shape is to be cut as well as the dimensions. The extent to which the manufactured shape meets its original design specification will largely reflect how this orientation was achieved as well as the quality of the plate from which the shape was made. Although it is true that many flame-cut parts are nothing more than 'space fillers' in fabricated structures, the example is not trivial for there have been instances for serious failures of stressed parts like reinforcing webs and fillets where a major structural failure can be traced back to unfavourable orientation in the flame-cutting operation.

Many materials are anisotropic, that is they possess different mechanical properties in different directions, and although this is well known for certain materials, such as timber where the effect of 'grain' is well recognized, it is not so well known that similar effects occur in metals and plastics. This position is not made any clearer by the fact that most commercial specifications include only properties in the most favourable direction. Designers and users must be aware that there can be quite significant property variations in different directions, and this applies just as much to substances such as high-impact polystyrene used for mouldings as to steels that featured in the above example. Figure 1.2 gives some idea of the order of magnitude of these differences. It is a saddening fact that, although such effects have been known practically as long as the material has existed, material specifications upon which designers base their calculations seldom set down lower limits for properties in the transverse direction.

The manufacturing specification usually consists of a drawing accompanied by detailed notes or instructions on how the article is to be produced. These instructions must not only be adequate to produce the shape and dimensional accuracy required, but must also specify items, such as times and temperatures to be used for intermediate annealing or to produce the final mechanical properties by heat treatment. It would be ridiculous to attempt to make every manufacturing specification fully comprehensive, say by specifying the ore body from which the metal must originate and detailing all the processes up to a finished product like a machine screw, so there are intermediate stages conventionally adopted as the starting points for a sequence of manufacturing operations. For example, a foundry would regard their starting material as the metal from a refinery and the scrap returned from their own processes or bought-in; a firm making large forgings would start with cast ingot, whereas one making small autombile forgings would regard their starting material as wrought bar. The starting point for small pressings would be thin sheet, and so on. Intermediate forms of material that are used as convenient starting points for subsequent manufacturing operations are referred to as 'semi-manufactured' products or 'semis'. The manufacturing specification thus begins with a full description of the appropriate semi-manufactured starting material and then goes on to give its own detailed description of process details for a particular end product.

As an illustration of the stages in a manufacturing specification, consider a small ball to be built into a ball and socket joint for use in the steering links of a motor car (Fig. 1.3). The designer has already anticipated the manufacturing operation in the sense that, in order to arrive at dimensions and shape of the part, he had to assume the properties of the forged medium carbon steel. This obviously is based on past experience of such parts as attempts to make them in other ways resulted in a heavier component or one that did not meet its intended function satisfactorily. The outcome of the design considerations is a drawing of the shape required for the ball pin such as illustrated in Fig. 1.3. The manufacturing specification begins by setting down the composition and size limits of the bar material to be used to make the forging. This has to be related to the forging reduction required to achieve the desired fibre structure and must state whether the part is to be forged as a whole or whether round bar stock is simply to be upset forged at one end to produce the ball shape. The temperatures and times for heating before forging must be laid down and there will almost certainly be an inspection of the rough forging before it is moved on to the next stage of the process. Heat treatment to produce the desired structure will require careful specification of time, temperature and atmosphere to prevent decarburization. Rough machining will be at

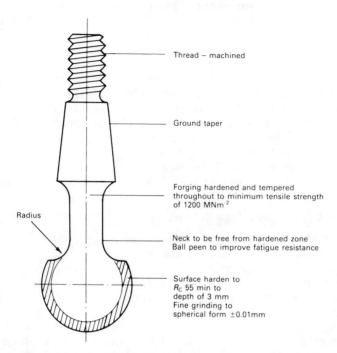

Thread – machined

Ground taper

Forging hardened and tempered
throughout to minimum tensile strength
of 1200 MNm^2

Radius

Neck to be free from hardened zone
Ball peen to improve fatigue resistance

Surface harden to
R_C 55 min to
depth of 3 mm
Fine grinding to
spherical form ±0.01mm

Fig. 1.3 Details of ball pin for ball joint

specified dimensions, leaving an allowance for the final grinding to exact size. The surfaces of the ball need to be locally hardened to reduce wear resistance, the most likely method being induction hardening, so the procedure to achieve the required depth of induction-hardened zone must be carefully controlled and samples will be required after the treatment for the hardness measurement and for checking the depth of the hardened zone. Great care must be taken at this stage to ensure that the hardened zone does not extend up into the neck of the ball joint or otherwise catastrophic failure may occur in service. The final manufacturing operations consist of grinding the ball and the taper to accurate dimensions, ball peening the shoulder and non-destructive inspection. As the component is to be used in the steering mechanism of a motor vehicle, then it would be wise to carry out 100% inspection for cracks before releasing from the factory.

The above example gives an indication of the many steps in manufacture that need to be specified even for a relatively simple component. No assembly or joining by welding was involved and the entire manufacturing operation was basically straightforward. Large, single components with detailed manufacturing procedures usually have a log that travels with the article through its entire manufacturing process and has to be signed at critical stages by a responsible person. This is not only to keep a record of the manufacture and to see that the specification is properly carried out, but such documents may become vital evidence in liability claims that may arise much later. One of the problems that industry is now having to face is that of storing these records for periods running into tens of years or, possibly, even the entire service lifetime of every article that is manufactured. Moreover, the engineer or 'person responsible' who signed that the quality and tests of manufacture were satisfactory may well be held personally responsible if a subsequent failure reveals negligence, whether at the design, materials specification, manufacturing or quality control stages. Hence, although it is clear that the manufacturing specification is essential in order to make an article, it is inextricably bound up with the quality assurance procedures and specification.

1.1.5 Quality control specification

At critical stages during the manufacture, checks need to be made that the product meets certain attributes of measurements laid down in the quality specification. Measurement of quality has been likened to measurement of the storminess of seas — easily recognized but difficult to quantify. 'Attributes' are features that can be assessed only by judgment, such as microstructure, surface finish, uniformity of appearance, etc., whereas things like composition, mechanical properties and dimensions can be measured and limits set on acceptable variations. All of the requirements for a particular product are set down in the quality specification, and, in order not to waste time and money in processing unsuitable items, checks are carried out frequently during manufacture so that products failing to meet the quality requirements may be rejected immediately their first shortcoming is discovered.

Although the customer stipulates what is and is not acceptable in the final article, the manufacturer lays down the quality control criteria throughout the production line. Although unsatisfactory products are rejected it does not follow that they will necessarily be scrapped, since repair or rectification of the quality fault may be possible. For example, it may be stipulated that, *inter alia*, a large steel casting must not contain any cavity larger than 6 mm in greatest dimension within the outermost 50 mm of the surface; in practice the time and cost of making such a casting and reaching the stage where defects of that nature can be found is so far advanced during the manufacture that it would be wasteful (and very costly) to scrap the casting. Accordingly, a repair by welding would be permitted, subject to another set of quality control measures. In a similar way, other items that fail a quality test may be recovered, whereas others may have to be rejected and scrapped, for example, a forging that has been overheated and the structure adversely affected, or a batch of small components that have been machined too small in a critical dimension.

Some of the detail of the quality control procedures will be dealt with in a later chapter, but one vital factor to bear in mind is that *quality costs money.* The more careful the quality control and the more frequently it is necessary during manufacture, the more costly it becomes and, of course, rejected items have to be paid for whether they are repaired or scrapped. At the end of the day it is the customer who must meet the bill, since no manufacturer can stay in business for long if his products are making a loss. Over-specification for quality can contribute to a commercial failure just as much as under-specification, though the latter is usually much more serious in terms of the consequences of parts failing in service.

Table 1.3 Example of a quality specification for a grey cast iron wormgear for outdoor exposure which must have a minimum tensile strength of 220 $MN^{-2}m$ (14 tons in $^{-2}$) and also requires a combination of wear resistance and machinability

Specification	Comments
(a) Grey cast iron (BS 1452: 1977 Grade 220)	Specification states that a **separately cast test bar** of the same iron will have the minimum tensile strength — an additional requirement might be that a test bar machined from the centre of the thickest part of a sample casting should have a minimum tensile strength of 220 MNm^{-2} (14 tons in^{-2}), or possibly that the entire casting be subjected to a fracture test. Material specification has clause calling for freedom from harmful defects, chill and carbides.
(b) Graphite type A size 4-5 (ASTM rating) in region of bearing surface Free Ferrite less than 5%	For optimum bearing properties and avoidance of scuffing.
(c) Critical sections, e.g. teeth and lugs must not contain graphite of type C or greater than size 3	To avoid fatigue failure
(d) 0.8% Copper	To stabilize pearlite in thick sections, thus ensuring that minimum strength level is achieved across the whole body of casting
(e) Phosphorus less than 0.2%	Necessary for application where additional embrittlement caused by phosphorus must be avoided, especially to resist shock loading on teeth. Also high phosphorus causes greater tool wear during machining
(f) Hardness 190-230 HB to be measured 1 mm below as cast surface in regions to be machined	To ensure homogenous machinability. To achieve this hardness it may be necessary to anneal the castings if surfaces have been chilled
(g) Dimensional tolerances	Specified on drawing, with due allowance for machining gear teeth
(h) Casting to be phosphated as final operation	Specify phosphating process. Necessary for running-in assistance, as well as a key and for corrosion protection under the paint film ultimately applied

A typical example showing how a basic material specification may have to be up-graded to meet a specific quality requirement is shown in Table 1.3, for a rack gear to be made in ordinary cast iron. The grade of cast iron specified represents about the commonest of the flake graphite irons, and the British Standard Specification gives little more than minimum tensile strength requirements, together with general clauses about freedom from harmful defects, chill and carbides (which would make machining difficult). However, to manufacture racks of good quality calls for much more than the minimum requirements, and Table 1.3 shows how composition and structure must be controlled to produce various attributes considered necessary for both successful manufacture and satisfactory service performance.

Inadequate specification of quality can have far reaching repercussions. For instance the cost of ownership is of vital concern to operators of capital intensive plants used in chemical processing, air and sea transport, oil-drilling platforms and the like, and trouble-free operation between the periods scheduled for maintenance shutdown would be part of any contractual specification. Hence any unexpected failure of a component due to inadequate quality or materials selection and/or design faults represents a substantial claim against the supplier. Even though the discovery that a part is defective after it is put into service may not result in a catastrophic failure, the monitoring of condition or de-rating of performance until it can be replaced can be a costly nuisance.

It is clear that quality control impinges upon the materials selection and manufacturing specifications, but it also stems from the market specification. To satisfy market requirements a product must be reliable and the only way to ensure this is to see that quality requirements are specified and properly carried out right through all stages of design and manufacture. If there has been one single issue that has caught the attention of designers during the last two decades, it is this aspect of designing-in quality at the outset. This means that, with complex structures, provision has to be made *at the design stage*, for the non-destructive inspection that needs to be carried out during manufacture. The days are gone when, at the time for the final inspection of, say, a pressure vessel, it is found that there is no access to the vital areas that need to be examined. In a more subtle way, the role of the metallurgist in ensuring that the structure of the material meets the desired standards after processing and heat treatment are just as vital in ensuring that the quality specification is properly satisfied.

1.2 EXAMPLES OF MATERIALS SELECTION CRITERIA

Although it is not the purpose of this book to deal in detail with selection of materials, it is worthwhile briefly examining the application of some of the principles referred to in the previous section, even if only to illustrate that each design must be considered in the light of its own individual demands. Both examples given below are simple and are intended to show

how the 'best' choice will alter when different design criteria are considered for the same application.

1.2.1 Selection for a tie bar

Let us suppose that several tie bars are needed for a particular application, each of which must be 2 m long, able to be threaded at both ends, and able to withstand a steady tensile force of 100 kN with a safety factor of 2. A short list of three different materials has been produced, each of which is available in suitable form and is able to fulfil all other requirements. Data for these three are set out in Table 1.4 and the problem is to select one of the three to satisfy in turn each of the following criteria:

(i)   if weight is to be kept to a minimum, regardless of cost; (i.e. as in an aerospace application);
(ii) if the requirement is to be met at minimum cost;
(iii) if the requirement is to be met with least elastic extension under the 100 kN force.

Table 1.4

Material	Density/kg m^{-3}	Elastic limit/MN m^{-2}	Tensile elastic modulus/GN m^{-2}	Cost as bar/£ kg^{-1}
Aluminium alloy	2700	332	71	4.50
Titanium alloy	4500	664	120	25.00
Alloy steel	7800	800	206	2.50

The first thing needed is the cross-sectional area of a bar in each material capable of supporting the tensile load of 100 kN with a safety factor of 2. The safety factor, in effect, halves the elastic limit of the material. Having established the cross-sectional area, the volume of a 2 m bar can then be calculated and hence its weight for (i) and cost for (ii).

Table 1.5

Material	Cross-sectional area/m^2		Volume of 2 m bar/m^3	Weight/kg	Cost/£
Aluminium alloy	$0.100 \times \dfrac{2}{332}$	$= 0.6024 \times 10^{-3}$	1.2048×10^{-3}	3.25	14.62
Titanium alloy	$0.100 \times \dfrac{2}{664}$	$= 0.3012 \times 10^{-3}$	0.6024×10^{-3}	2.71	67.75
Alloy steel	$0.100 \times \dfrac{2}{800}$	$= 0.0250 \times 10^{-3}$	0.5×10^{-3}	3.90	9.75

The calculations are shown in Table 1.5. The choice for the first two criteria is thus clear; the titanium alloy offers the least weight and the alloy steel the lowest cost.

Table 1.6

Material	Stress in bar 0.5 elastic limit/GN	Modulus/GN m^{-2}	Extension/m
Aluminium alloy	166×10^{-3}	71	2.33×10^{-3}
Titanium alloy	332×10^{-3}	120	2.76×10^{-3}
Alloy	400×10^{-3}	206	1.94×10^{-3}

For the mechanical requirement of least elastic extension under the force, it is necessary to use the tensile modulus at the stress acting, which we know from the first set of calculations is equal to half the elastic limit. The calculations are given in Table 1.6, from which it can be seen that the alloy steel gives the least extension. Obviously different design criteria could be adopted and many more materials included in the short-list, but the point is clear that the 'best' material choice is sensitive to the selection criterion. Other considerations would clearly need to be taken into account in a real situation, since properties such as toughness, corrosion resistance and environmental and manufacturing factors were ignored in this example.

1.2.2 Selection for electrical conductivity

Metals are the only materials that are useful as electrical conductors and some are much better, relatively speaking, than others. Table 1.7 gives the electrical resistivity of most of the common metals which could be made as wires or strips suitable for electrical conductors.

Alloying invariably raises the resistivity of pure metals as exemplified by the values for iron and stainless steel, aluminium and aluminium alloy, and copper, brass and Monel®. This is true in all cases, so if electrical conductivity is paramount, then the metal must be pure or, at most, a very dilute alloy.

Silver is better than copper, but only marginally so, and it is much scarcer and more expensive. During the early years of electricity copper reigned supreme, but aluminium has long substituted for copper in certain electrical applications, but not all. In certain kinds of electrical machines it is necessary to pack as many conducting elements as possible into the space available, so the best performance can only be obtained from the wire with the *lowest resistivity per unit of cross-sectional area*: from silver (best), or from copper (marginally second best). As the data in the fifth column of Table 1.7 demonstrate, aluminium wire would have a 50% larger cross-section than copper to give the same electrical resistance.

Table 1.7 Parameters to be considered in selecting a material for an electrical conductor

Material	Density (kg m^{-3})	Electrical resistivity (10^{-9} Ωm)	Relative conductivity (copper = 100)	Cross-section with same electrical resistance as 100 mm^2 of copper	Resistivity × density (10^3 kgm^{-2}Ω)	Weight of conductor having same resistance as copper per unit length (copper = 1.00)
Copper	8 900	1.67	100	100	15.0	1.00
Brass (70 Cu–30 Zn)	8 520	6.33	26	384	53.8	3.59
Monel (70 Ni–30 Cu)	8 800	48.0	3	3 333	422	28.1
Nickel	8 900	6.84	24	410	60.9	4.06
Iron	7 870	9.71	17	581	76.4	5.09
Stainless steel	7 930	70.0	2.4	4 166	555	37.0
Lead	11 340	20.66	8	1 237	234	15.6
Zinc	7 140	5.91	28	354	42.2	2.81
Aluminium	2 700	2.65	63	159	7.16	0.48
Magnesium	2 800	5.0	33	303	14.0	0.93
Aluminium alloy	1 740	4.45	38	266	7.74	0.52
Gold	19 300	2.42	69	145	46.7	3.11
Silver	10 500	1.62	103	97	17.0	1.13
Sodium	970	4.5	37	269	4.37	0.29
Carbon	2 300	1375	0.12	82 335	3163	211

Hence, per unit of volume, copper is considerably superior to aluminium.

However, there are many instances where obtaining the minimum cross-sectional area of conductor is not important, e.g. for overhead power transmission lines where there is plenty of space available and the criterion is to secure *maximum conductivity per unit of weight* of conductor rather than per *unit of volume*. In terms of being able to transmit electrical current with least energy loss, aluminium is actually over twice as good on a weight-for-weight basis as copper. If, by adopting different criteria, the ranking of these two metals can change, then it is worthwhile looking at the other metallic materials. The two columns at the extreme right of Table 1.7 indicate the ranking in terms of *resistance per unit of weight*. In the extreme right-hand column, any metal with a value of less than 1.0 is better than copper; it has a lower resistance per unit weight, despite requiring a larger cross-section than copper. You will see that aluminium is, in fact, the best conventional material, with magnesium running a close second. Copper is third best, but the interesting feature is that sodium is better by far than all the others. Now sodium is a highly reactive metal — it explodes on contact with water. Nevertheless, might not conductors be made of sodium encapsulated in some kind of protective sheath? They would be nearly four times better than copper and twice as good as aluminium. The technical problems appear daunting, but are not unlike those that had to be overcome with titanium, and even before that, with aluminium before they took their place as useful engineering materials. In this situation, however, before any reader tries to exploit this potential of sodium he must not lose sight of the fact that superconducting materials are already well developed and are much better than sodium, provided that they are operated at a temperature below about 30K. These represent a totally different solution to the problem of efficiently carrying electrical energy.

The substitution of aluminium for copper for overhead power transmission lines has still not been completely justified. As indicated earlier, alloying drastically reduces conductivity, yet is one of the most important ways of strengthening metals. Pure metals may be best for electrical conductivity, but are weakest in terms of yield stress and tensile strength. Electrical conductors in overhead transmission lines need to be suspended between supports spaced as far apart as possible, but to do this a high yield stress is desirable. Pure copper is roughly twice as strong as pure aluminium, so it now appears attractive since its resistivity per unit of *mechanical strength* is marginally better than aluminium. However, in practice the mechanical weakness of aluminium is overcome by supporting it by means of high tensile steel cable, usually running up the centre of the conductor, so that the aluminium does not have to support any of its own weight. The same could be done with copper, of course, but the weight factor would still apply in the same way, leaving aluminium twice as effective. This is the reason why copper is not used for power transmission lines, bus bars and in certain kinds of electrical devices and machines. However, there are exceptional applications where frictional heating effects are likely to be encountered, for example overhead power lines for electric railways, which still utilize a dilute copper–chromium alloy because aluminium cannot withstand the service conditions involving friction and arcing.

When electricity reaches the home, the picture changes. In some countries, aluminium wire is used for domestic wiring and distribution, but this has not yet occurred in the United Kingdom or other European countries because there are serious problems with such systems. The low mechanical strength of aluminium is one factor, since it tends to be easily broken during fitting (e.g., by screwing a contact down too tightly). Its tendency to form an oxide layer at the surface is a nuisance since the oxide is a good electrical insulator; poor electrical contacts become heated, oxidize faster and eventually cause a break in the circuit. Joining of aluminium wires is difficult for the same reason. Although aluminium can be soldered to make joints, it is not an easy technique, and the solders are very prone to corrosion in moist atmosphere, whereas copper is one of the easiest metals to join by soldering and joints last a long time. So, on balance, copper is still the most reliable conductor for domestic wiring.

It will be realized that this second illustration of selecting a material involves a completely different approach from the tie bars. In both cases the short-list was easily drawn up by the kind of criteria referred to in Section 1.2.3, but the final choice could not be made without detailed consideration of the requirements for a specific application. The second example also illustrated how a totally different solution to a problem may be arrived at by a different discipline. Instead of trying to find a strong conducting material on the basis of material properties alone, the solution to the problem of strength of the conductors could be mechanical, i.e. supporting a good conductor with a steel core; or electrical, reducing I^2R loss by increasing voltage and by keeping the conductor cool (to minimize the resistance, R), or to use superconducting cables cooled by liquid helium.

Such diverse solutions illustrate the interplay between designer and materials scientist, as well as highlighting the difficulty of generalizing about selection of a material.

1.3 RESOURCE AND ENERGY IMPLICATIONS
OF MATERIALS SELECTION

All the physical sources from which engineering materials are derived are in the outermost 3 km or so of the earth's crust. This is now sufficiently well explored for it to be recognized that there are finite limits to the amount of recoverable resources and some of these are fast being approached, judging by the rates of consumption over the last 50 years. Others, fortunately including iron and aluminium, are sufficient to last for many hundreds of years. Market price reflects supply and demand, but with scarce resources no amount of market pressure can produce a commodity when there is none left to extract. In such an event, society would have to manage by re-cycling that which is already in use and, to distinguish between the two sources of supply, a metal that is extracted from an ore and being used for the first time is referred to as 'primary' whereas one that has been recovered and re-cycled is known as 'secondary'. The problem with all physical resources is that it is impossible to know exactly how much there is available and therefore to predict when it is likely to be consumed, if at all. Metaliferous minerals are present in many rocks and when the amount is sufficient to justify extraction at current market prices then the particular rock is known as an ore. Hence deposits whose composition is currently below the minimum ore grade may well become profitable to treat if the market price rises, and this could bring into circulation considerable extra amounts of the scarce resource.

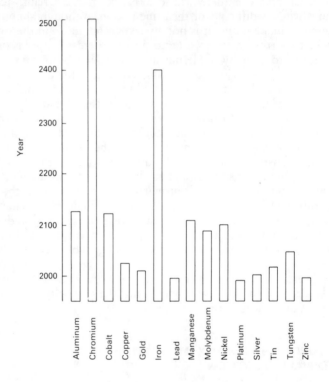

Fig. 1.4 Lifetimes of present reserves for various metals at 1980 extrapolations of rates of extraction

The above makes it clear why predictions such as those given in Fig. 1.4 are always open to question and are continually being revised. Neverthe-less, they do serve to highlight metals that are likely to have supply difficulties in the near future and hence have market prices that are likely to rise steeply in comparison with others. Everyone concerned, therefore, with designing and selecting materials should bear in mind these possible future limitations. For example, the essential ingredients of tool steels include tungsten, chromium, vanadium, cobalt and molybdenum; some are already in the situation of having resource problems in the near future and it is therefore advisable to conserve them either by substitution or other alloying elements (e.g. molybdenum for tungsten) or by making tools with inserts so that only the portion that does the cutting is made of the particular alloy while the remainder of the body of the tool is composed of steel with lower amounts or different alloying elements.

Substitution is one answer to conserving a scarce resource metal, but it is by no means possible in every instance. The chemical or physical properties may be dominant for a particular application and there may be no other element or material that can effectively substitute. For example, the chemical properties of platinum, palladium and vanadium as catalysts are unique, and tungsten (which has the highest melting point of all the metals) may not be substituted effectively in certain applications by anything with a lower melting temperature. If tungsten ceased to be available for lamp filaments then a radical redesign would be necessary to meet the need for illuminating devices.

Recycling metal from machines that have outlived their usefulness is commonly believed to be the answer to resource limitations, but this secondary supply can, at best, merely serve to extend the life of a resource by a modest period of up to a few hundred years. Secondary metals are derived from two sources, process scrap and 'old' scrap, the former being of known composition and quality whereas the composition and quality of latter can be uncertain and so it can present more problems of recycling, to say nothing of dismantling old machines, ships, buildings, etc. The present levels of recycling of the non-ferrous metals (whose value makes them much more attractive for recovery than ferrous) are in the region of 40%. This could be raised if economics justified the extra cost when a resource became very scarce, but some uses are dissipative so recovery could never approach 100%. Figure 1.5 shows how the useful life of a hypothetical metal resource would be extended by a mere 100 years if the recycling level was 50% after an average service lifetime of 25 years. During the period after the primary resource became exhausted, the amount of metal available would be drastically reduced. Even if recycling could be 100%, the best that could be hoped for would be a constant amount of metal in circulation, most of which would be 'locked-up' for the lifetime of the artefacts in which it was used.

The sea is an enormous potential source of most of our metals, but the great dilution makes their extraction economically impracticable at today's prices and with today's technology. Many rocks, too, contain minute amounts of the useful metals, but again their recovery is impracticable. However, even if the technology became available to treat these sources and assuming that the political, social, economic and geographical problems could be overcome, there is another major difficulty influencing their extraction, namely that of energy. At present copper is extracted from ores containing something below 0.5% which means that to extract 1 tonne of metal 200 tonnes of ore has to be mined and treated. To reduce

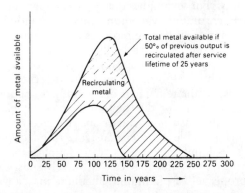

Fig. 1.5 Effect of recycling on extending the lifetime of a metal resource. (At every 25-year interval, half the *total* metal available 25 years earlier is assumed to become recycled and added to primary production. Notice how when once the primary source is exhausted the quantity available falls steeply)

the ore grade to 0.1% — and there could be immense potential reserves of copper at this level — would involve the treatment of 1000 tonnes of rock for every tonne of metal extracted. The colossal scale of such mining operations would be staggering to meet a growing future world demand, but more importantly the energy required would be prohibitive.

Energy is the lifeblood of technology-based societies; in 1981 roughly 15% of the world's energy consumption was used on extracting metals and making artificial engineering materials (cement, glass, etc.). With increasing demands from agriculture for fertilizers and machines, for desalination and food processing to support the growing population, it is difficult to see how any large increase in energy demand for metals could be met.

Table 1.8 lists present-day energy consumed in producing various materials. The trend in all cases will be upward to reflect the increasing cost of energy, and it is clear that the energy-hungry metals, such as aluminium and titanium, will increase more steeply than less demanding ones. In turn this cost will follow through into finished products and it may well be that the basis of materials selection in the not too distant future will be total energy content in relation to the technical requirements of the application.

Wise choice of materials will therefore take into account questions of availability and energy content, as well as designing with an eye to facilitating the recycling of component metals (see Chapter 7).

1.4 SPECIFICATIONS AND BRAND NAMES

In recent years efforts have been made particularly in Europe but also

Table 1.8 Energy consumed in producing different materials

Material and product form	Specific energy content ($KWh\ kg^{-1}$) Primary source	Secondary source (additional energy)	Approximate energy content per MNm^{-2} of property Tensile strength (KWh per MNm^{-2})	Elastic modulus (E) (KWh per MNm^{-2})
Cast iron (casting)	14	5	100–300	0.7–2.1
Steel				
mild steel bar	16	6	350	1.6
strong alloy steel bar	18	6	125	1.7
stainless steel sheet	30	6	230	2.9
Aluminium				
pure, sheet	80	12	750	8.4
alloy extruded bar	80	—	430	8.4
Copper				
domestic water pipe	30	8	440	4.2
Magnesium				
extruded bar	115		1000	11.1
Titanium				
alloy bar	155		700	15.2
Plastics				
polyethylene film	30		1800	300
nylon rod	50		700	20
PVC tube	50		850	25
Timber				
softwood	0.5		50	0.14
hardwood	0.5		25	0.07
Reinforced concrete beam	3.5		200	0.85

Energy consumed in producing 1000 of 25 mm bore pipe (MWh)

Iron (galvanized)	2815
Copper	800
Polyethylene	695

worldwide to establish specifications so that metals and alloys made in different countries are completely interchangeable and replaceable, that is they have the same combination of mechanical and physical properties.

Most of the tonnage metals and alloys produced by different manufacturers are now freely interchangeable, and tables are available showing the equivalence of these materials as specified in the different countries. There has also developed a series of European specifications known as *Euronorm* and a worldwide set of common specifications known as ISO (International Standards Organization). The tendency in both these series is to emphasize more strongly the minimum mechanical and physical properties required, and to reduce the importance of meeting the chemical composition and impurity limits in all cases where they are not likely to be of significance in service.

For two of the main alloy groups, namely aluminium and stainless steel, the American specifications have been adopted by most industrial countries. Also because of the American lead in mass production of cars, many of the mechanical engineering specifications are in SAE (Society of Automotive Engineers) terms. A similar state of affairs exists for the chemical and petrochemical industries where API (American Petroleum Institute) or ASTM (American Society for Testing and Materials) are dominant.

At the other extreme, and mainly for the low tonnage and specialist uses and applications, engineers will find that metals and alloys are still regularly and almost universally specified according to brand name, product or alloy name. Examples of such uses are: Condenser Tubes, Die Steels whether hot or cold, Cutting Tools, Tungsten Carbide Cermets, Jet Turbine alloys. In these cases, specifications may be developed jointly by the producer and his individual customers, but great care must be taken when switching from one brand name to another.

CHAPTER 2

The origins of the strength of metals

The data concerning the properties of a metal that must be considered during the design process will be presented as sets of figures or statements that describe the mechanical, physical and chemical properties of a specific material in a specific condition. To use such data to best advantage it is necessary to understand the properties from which materials derive their strength and how these can be varied through changes in composition and production processes.

Materials derive their properties from:

(i) the interaction of the atoms from which they formed
(ii) the behaviour of *groups* of those atoms (which may or may not have a regular, crystalline structure)
(iii) the attributes associated with *assemblies of those groups of atoms.*

Thus, to provide an understanding of the properties of materials this chapter will discuss the influence of atomic and crystalline structure as well as the behaviour of materials in bulk form.

2.1 THE NATURE OF BONDING IN SOLIDS

The most useful model of atomic structure is to regard an atom as a positively charged nucleus containing the atom's mass, surrounded by virtually massless electrons sufficient in number to balance the positive charges on the nucleus, leaving the atom as a whole electrically neutral. The electrons are arranged in various energy bands and the outermost band is the most loosely bound to the nucleus. Figure 2.1(a) shows two-dimensional representations of magnesium, a typical metal atom, which has two electrons in its outermost band, and oxygen, a typical non-metal, which has six electrons in that band. Both have two electrons in their innermost band, nearest the nucleus.

Atoms appear to lose their impetus to interact when they have eight electrons occupying their outermost band. Those that do not have this configuration (which is possessed only by the few inert gases) are always trying to associate themselves in ways that approximate to their achieving it. It is this characteristic that gives rise to the three forms of atomic bond: *ionic, covalent,* and *metallic* bonds.

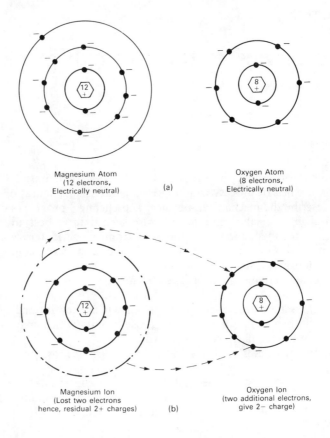

Fig. 2.1 (a) Isolated atoms of magnesium and oxygen showing arrangement of electrons in bands. (b) capture of two electrons by oxygen produces charged ions

2.1.1 Ionic bonds

Ionic bonds occur between the atoms of metals and non-metals and are very strong. The resulting material is characterized by a high melting temperature, hardness and brittleness. A typical ionic bond is formed when an atom of oxygen takes the two outer electrons from a magnesium atom (Fig. 2.1(b). In doing so, the oxygen atom gains two negative charges and the magnesium atom, deprived of its two outermost electrons, is left with two positive charges. Both the oxygen and the magnesium now have eight electrons in their outermost rings and, like the inert gases, are chemically 'satisfied'. However, the two atoms which were electrically neutral have now developed opposite electrostatic charges and this causes them to bond together, as shown in Fig. 2.2(a). This figure is a two-dimensional representation of the compound magnesium oxide (MgO).

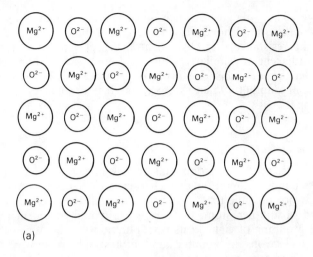

(a)

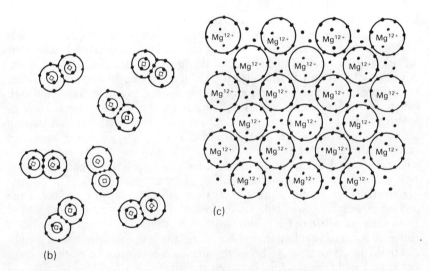

(b)

(c)

Fig. 2.2 The three different types of bonding in materials

Atoms with similar charges repel, whereas atoms with opposite charges attract each other. Thus, in bulk material consisting of ionically bonded atoms, a regular pattern called a *crystal structure* is formed which is three-dimensional and in which each atom is surrounded by atoms carrying the opposite charge. The strength of such materials is explained by the strength of the electrostatic bond between the dissimilar atoms, and the brittleness by the resistance that charged atoms display to being forced into positions adjacent to similarly charged atoms. Magnesium oxide resists forces that would otherwise bring oxygen atoms or magnesium atoms next to one another. When the force is great enough, the crystal simply breaks apart.

2.1.2 Covalent bonds

Covalent bonds occur between atoms that have four or more electrons in their outermost band, a condition that characterizes elements known as *non-metals*. It is not possible for one atom to take all the elecrons in the outermost band of another atom. If it did, there would be too many to make the ideal grouping of eight in its own outermost band. When there are four or more electrons in the outer band, atoms associate in ways that enable them to *share* their outermost electrons, as indicated in Fig. 2.2(b). This Figure shows two oxygen atoms sharing electrons so that each has the preferred eight electrons. The bonds between the atom parts are very strong, but those between the individual pairings are weak; so weak in fact, that oxygen does not solidify until very low temperatures are reached, when it will then form a crystal structure.

Covalently bonded substances range from gases through liquids to solids and all covalent bonds tend to be very strong. For engineering applications we are more interested in solids, and the most relevant example is carbon. An atom of carbon has four electrons in its outermost band and, in order to achieve the preferred eight electrons may, for example, associate with other carbon atoms or with four single-electron-sharing atoms such as hydrogen to form methane (CH_4), or with two twin-electron-sharing atoms such as oxygen to form carbon dioxide (CO_2). When it associates solely with other carbon atoms two crystalline carbon forms are possible. The first form has carbon atoms in a tetragonally linked cubic structure, whereas the second has them linked in hexagonal plates. The former arrangement gives diamond, the hardest substance known, whereas the second is graphite, which has well known lubricating properties derived from the plates being able to slide over each other. This structure involves a different form of bonding that although resulting in each atom having eight electrons in its outermost bond the distance between the plates is large compared with the distance between the atoms that form each plate and consequently the binding force between each plate is weak. Additionally this form of bond uses up three electrons per atom and the fourth is free or mobile in a plane parallel to the layers.

Carbon atoms covalently bonded to other atoms such as hydrogen frequently link together to form long chains. Covalent bonding between the atoms of these chains or *polymer* structures is not always reflected in their properties because although the chains are strong they are also flexible and the bonding between neighbouring chains is weak.

2.1.3 Metallic bonds

Two thirds of all the elements have fewer than four electrons in their outermost bands. Although these electrons are sufficient to balance the positive charges on the nuclei, if two such elements form a bond, the number of electrons in their outermost bands is insufficient, even if shared, to provide the preferred eight electrons in those bands and for ionic or covalent bonding to occur. In the solid state these elements form an entirely different type of bond, which characterizes substances known as metals. The electrons in the outermost bands of all the atoms in a metal drift in clouds through the spaces between the positively charged nuclei plus other electron shells as illustrated in Fig. 2.2(c) for magnesium.

The nuclei and their inner bands of electrons act as if they were hard balls and packed closely together in a regular pattern to take up what is called a *crystalline array*. Representation of such arrangements common in metals are shown in Fig. 2.2(c) it is easy to visualize layer upon layer of similar arrays packed closely together to form a three-dimensional crystal *lattice*, such as actually occurs in practice. This array of positively charged ions is held together by its attraction to the cloud of negatively charged electrons, a unique form of bonding called *metallic bonding*. Because no ion has any particular preference for being in a particular position, ions can be moved about within the space lattice without disturbing the regularity. Furthermore, the cloud of electrons can be made to drift in a given direction under the influence of an applied electrical potential, providing an *electric current*. Electrical conductivity is typically another unique characteristic of metals. In ionic or covalently bonded crystals the electrons are involved in the bonding and are not free to drift, or to flow under the influence of small electric potentials; only when extremely high potentials (called breakdown potentials) are applied can the electrons be torn away.

2.2 BONDING AND THE EFFECT OF EXTERNALLY APPLIED FORCES

In addition to the comparative freedom of electrons in metallic bonds, another major difference between metallic and the other types of bond is their behaviour under the action of externally applied forces. Small forces only slightly stretch or contract the bonds in all three types, which recover when the force is released. This property is called *elastic extension* or *elastic compression*. However, when larger forces are applied, metal-bonded ions ride over each other to form a similar pattern that persists after the force is released. This is possible because all the ions are similar in nature and the electrons are not tied to any particular atom. In contrast, ionically bonded atoms resist such sliding because they and their electrons are tightly bound. Consequently, ionically bonded materials tend to be brittle.

Because of the ability of their nuclei to ride over each other, metallically bonded crystals can be shaped using mechanical forces and the bonds

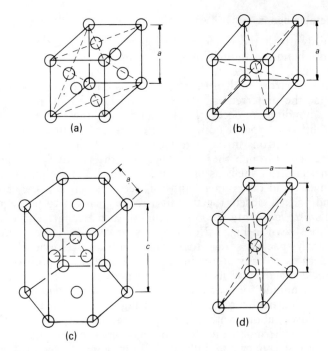

Fig. 2.3 Space lattices of crystal structures found in metals. (a) Face centred cubic; (b) body centred cubic; (c) close packed hexagonal; (d) body centred tetragonal

between the atoms remain as strong. This property is known as *ductility* or *malleability* and is characteristic of the metallic state.

Whatever the form of bonding, most substances form regular three-dimensional arrays (or crystal structures) in space. There are fourteen such structures, but of these only four are commonly found in metals used in engineering applications. These are illustrated in Fig. 2.3, as *space lattices*, i.e., the simplest single cell representing the very large numbers of atoms in the three-dimensional array in the bulk crystal.

If one considers the millions of atoms that must associate to form a crystal it is not surprising that some imperfections and irregularities occur. It is these defects that determine the bulk properties of a material.

2.3 THEORETICAL AND ACTUAL STRENGTH

Whatever the crystal structure, bonds between atoms act like springs in the way they react to externally applied forces. If we consider a perfect crystal (i.e., one in which there are no displaced atoms or irregularities), the externally applied force (stress) will displace all the atoms in one plane with respect to those in the adjacent plane, causing all the bonds to stretch to the same extent. This displacement is called *strain*. When the force is

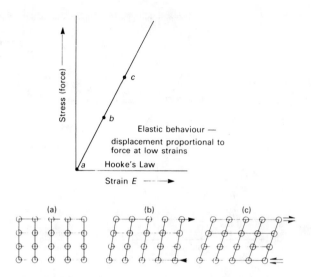

Fig. 2.4 The early part of the stress–strain curve in relation to the displacement of atoms in the crystal structure

removed, the atoms return to their original positions. This behaviour when represented on a graph showing the relationship between the applied force and the consequent stretch or strain, is the first, straight part of the stress–strain curve (see Fig. 2.4). It was Hooke who stated in 1679, 'As the extension, so the force', though he was concerned with wires rather than atoms.

This elastic straining, when a material obeys Hooke's Law, should (but in practice, does not) continue to the point where the atoms are displaced nearly halfway to their next stable position. Obviously, if they were strained precisely to the halfway position (as shown in Fig. 2.5) there would be as strong an attraction to the next rest position as a restoring force back to the original one. In other words, true elastic behaviour might be expected up to strains of about one third of the atomic displacement. This, however, is much greater than elastic strains observed in practice in any actual material, which seldom exceed 1% and are frequently much less even than this.

If reversible elastic behaviour was exhibited by bulk materials even up to strains of only 10–15% those materials would show yield strengths between 10 and 100 times greater than those found in practice. The explanation of why bulk materials do not exhibit strengths even approaching these values lies in the irregularities that occur in their crystal structure. The two most important irregularities that determine the properties of strength and toughness are *dislocations* and *micro cracks*, which exist in all bulk materials.

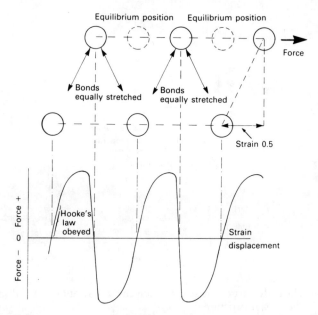

Fig. 2.5 Showing how force to displace atoms increases with displacement to some position near $\frac{1}{3}$ of lattice spacing; thereafter attraction towards the next position causes departure from linearity. At exactly half way the pull back to the original position is balanced by that towards the next position, so the displacing force becomes zero

2.4 DISLOCATIONS

Dislocations are faults in the crystal structure of a planar character in which an extra half-plane is inserted or is missing inside the crystal structure. It gives rise to a local distortion centred on the end of the missing or additional plane. It was shown in connection with Fig. 2.5 that if an applied force moved a plane of atoms to a position midway between the two stable positions, the atoms could move to the next rest position or back to the previous one with equal ease. So, if the atoms were balanced at their mid-positions, a tiny applied force would be sufficient to start them forward to the next or back to their original position.

Consider now what would happen if, because of an irregularity in the crystal structure most of the atoms were in stable lattice positions but a few were midway and, therefore, in unstable positions [see (Fig. 2.6a); there is an extra atom in the top row]. If a small force is applied to this structure the behaviour would at first be elastic and obey Hooke's Law. But if a larger force is applied the atoms that are in an unstable position will slide to stable positions, restoring the regularity of the structure in their neighbourhoods. However, further along, some atoms previously in *stable* positions will have moved to *unstable* positions (Fig. 2.6b). The

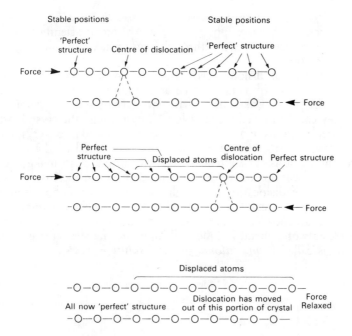

Fig. 2.6 Schematic diagram showing how a dislocation can move under applied force and allow movement of atoms one row at a time

overall effect is to move the irregularity through the crystal in the direction of the applied force (Fig. 2.6c). It will be clear that it takes a smaller force to achieve this movement in an irregular array than would be required if all the atoms were in stable positions, since in the latter case the applied force would have to move every atom at the same time.

In Figure 2.6 the two lines of atoms illustrated are cross-sections of two planes of atoms lying vertical to the plane of the paper and the irregularity is a cross-section of a line of irregularities that is vertical to the plane of the paper. If a force is now applied to the planes of atoms the line of irregularities will move through them in the form of a linear defect. This distorted region is called the *dislocation*, and the linear form just described is only one of several configurations which dislocations can assume. The crystal geometry can be very complex, but the essential idea of a few atoms being able to shear at any one time under the influence of a small applied force is simple, yet it explains why no bulk material can ever achieve its theoretical strength level.

Dislocations are a natural part of all crystals and if a metal has not been subjected to mechanical deformation, there are about 10^6 dislocation lines intersecting each square centimetre of surface. If this metal is mechanically deformed, this number is multiplied by interactions to around 10^{12} dislocations per square centimetre. In most metal crystals, dislocations move when an applied force reaches a sufficiently high level and, by the very act of moving through the crystal, cause it to deform rather than fracture.

The importance of dislocations cannot be overemphasized, since all mechanical behaviour in metals can be accounted for in terms of whether dislocations are *pinned* (fixed in a particular place by some mechanism) or whether they can be moved more of less freely through the crystal lattice. Thus, if 'foreign' atoms of a different size are introduced into a crystal they will cause lattice distortions in addition to those caused by the naturally occurring dislocations. The effect of such distortions is to make it more difficult for dislocations to move through them and to increase the force required to move dislocations through the lattice, with the result that elastic behaviour is retained to a higher stress level. The material is therefore strengthened. The introduction of atoms of different elements into the crystals of a metal is called *alloying* and the strengthening effect that results is called *compositional* or *solid solution hardening*.

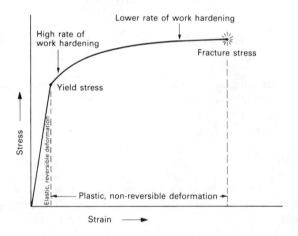

Fig. 2.7 Stress–strain curve for a ductile metal below $0.6T_m$

A further example of the importance of dislocations is the *work hardening* that occurs when a metal is subjected to *plastic deformation*. In the stress–strain diagram (Fig. 2.7), plastic deformation begins where the straight-line stress–strain relationship ends. In other words, a plastically deformed metal will no longer revert to its original form when the applied stress is removed. Plastic deformation occurs by dislocations moving and multiplying through the crystal lattice. This multiplication adds greatly to the irregularities in, and distortion of, the crystal and makes it more difficult for further movement to take place,. As plastic deformation increases, so the level of applied stress needed to force dislocations to move increases. The practical effect is that metals become harder and stronger when deformed. In other words, they *work harden*.

The effect of temperature on dislocations is also important. At high temperatures, diffusion causes atoms to move more rapidly and such movement within the crystal lattice can destroy dislocations and reduce the tangles that have built up during deformation. Hence, after a period of heat-treatment which reduces the number of dislocations, a smaller stress is sufficient to start the remaining dislocations moving again. It is this mechanism that is involved when a metal that has been work hardened is heated to become softer and more ductile, a practice known as *annealing*.

2.5 MICRO CRACKS

Dislocations are not always mobile and in certain metallic crystals, and frequently in ionically and covalently bonded crystals, they remain fixed even when high levels of stress are applied. However, when dislocations are firmly anchored, such materials still do not realize their theoretical strength because of other irregularities in the crystals of bulk materials called *micro cracks*. Whether a bulk material has a crystalline structure, is *amorphous* (has no recognizable structure at all), or is a mixture of the two, it contains micro cracks. These can be regarded as discontinuities lying in one plane which range in size from a few tens of atoms up to cracks visible under a microscope. Micro cracks concentrate stresses at their tips, and grow when subjected to much lower stresses than those needed to distort or fracture unblemished material.

Micro cracks exist in all solids, but in metals their response to an externally applied force is offset by the presence of mobile dislocations close to the crack tips. Usually, these dislocations move under the applied force, producing plastic deformation and relieving the stresses at the crack tips, thus preventing the cracks from growing. However, when dislocations are not able to move, cracks do propagate and failure quickly follows. Failure in these cases is brittle, i.e., exhibiting no plastic deformation, and occurs while the material is still behaving elastically.

A material does not always fail when put under a stress sufficient to propagate micro cracks. If there is nothing to stop the cracks developing they will indeed spread rapidly throughout the section and brittle failure will follow. However, other interfaces can exist within the material and these may not be orientated in directions favourable for further growth. If a crack reaches one of these it cannot continue to propagate on the other side of the interface. Instead, it will change direction, as indicated in Fig 2.8.

Such interfaces may be introduced deliberately into materials subject to brittle fracture. For example, many composite materials, both artificial, such as glass-fibre-reinforced plastic, or natural, such as timber, contain very many interfaces that stop the growth of cracks running into them from a roughly perpendicular direction. Preventing the propagation of micro cracks is an effective method of increasing the elastic strength and toughness of a material.

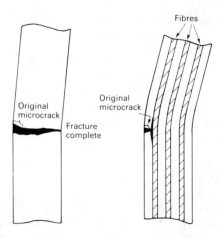

Fig. 2.8 How introduction of interfaces by fibre strengthening prevents catastrophic propagation of microcrack

If a material could be produced entirely free from micro cracks, or with just a few micro cracks aligned parallel with the direction of the applied tensile force, it would have an elastic strength close to the theoretical value. These conditions do prevail in newly produced thin fibres (of the order of 10^{-4} mm diameter). Unfortunately, micro cracks soon start to appear and the strength of the fibres falls with time; it also falls as the thickness is increased, due to the greater probability of a crack being present oriented in a direction in which it can propagate under external forces.

When sufficient stress is applied to a material it is to be expected that micro cracks normal to the direction of tensile stress will propagate rapidly and failure occurs in the elastic region, which is what happens with brittle materials in bulk form. However, in metals it is dislocation movement in bulk materials that dominates the behaviour under stress and instead of suddenly cracking, the material yields and begins to deform as atoms slip over one another. As the stress is increased, deformation continues and dislocations interact to form tangles. These not only make it more difficult for further movement, they also create so much localized distortion that they, in effect, constitute a large (atomically speaking) crack.

Such deformation cannot continue indefinitely and ultimately the cracks link up and fracture occurs as indicated in Fig. 2.7.

The picture is rather different when dislocations are pinned by alloying, or in some other way are prevented from moving under applied stress. In extreme cases when such pinning of dislocations has occurred metal fracture can take place without plastic deformation in a manner similar to that described earlier for brittle solids.

2.6 VACANCIES

Dislocations and micro cracks are not the only imperfections commonly found in a crystal lattice. Occasionally, positions in the lattice which should be occupied by atoms are not filled. Such sites are known as *vacancies*. Although vacancies do not have a major effect on mechanical properties they are important in determining the behaviour of solids at high temperatures, or when atoms diffuse through the lattice. At every temperature there is an equilibrium number of vacancies necessary for the crystal to be thermodynamically stable; just below the melting temperature the proportion is of the order of 10^{-3}, which means there is one vacancy in every cube of 10 atoms $\times$ 10 atoms $\times$ 10 atoms.

Diffusion is important in heat treatment, the sintering of metal powders, in the surface-hardening of metals, and in other manufacturing processes. All these depend on the movement of 'foreign' atoms through the crystal lattice. The number of vacancies in a lattice increases exponentially as the melting-point is approached. Consequently, the above processes tend to be carried out at high temperatures when the number of vacancies is large and the diffusion rate is rapid.

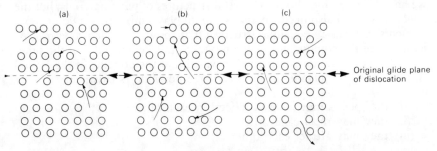

Fig. 2.9 How vacancies diffusing through lattice may 'eat away' atoms and cause a dislocation to climb out of its original plane

Vacancies are able to absorb dislocations within crystals and eliminate some of the tangles that occur during extensive deformation. Figure 2.9 shows that, as vacancies 'flow' through the lattice, they 'eat away' the end of the extra plane forming the dislocation; the dislocation 'climbs' out of its original plane. Subsequently, it can no longer act as a barrier to the movement of other dislocations on that plane, so further deformation is facilitated.

Vacancy flow is one of the processes that takes place when cold-worked metals are annealed, and as explained earlier, the consequent elimination of dislocation tangles makes the material softer, lowers its strength and makes it able to withstand a further amount of plastic deformation.

The flow of vacancies to dislocations is also partly responsible for *creep* in metals. Creep is the plastic deformation of metals that can occur when they are subjected to a constant stress at a level much lower than that which will lead to fracture if the stress is applied quickly. When a stress within the elastic region is applied, several dislocations are strained to the point where they are just about to move. If, in this condition, vacancies diffuse to the ends of dislocation lines causing them to climb out of their original planes, they no longer act as barriers, and over a period of time, extensive movement of dislocations will cause permanent deformation or creep. Slow deformation occurs under stresses much smaller than those indicated by the linear portion of the stress — strain curve shown in Fig. 2.7. This deformation is therefore plastic and time-dependent. Because there are many more vacancies in the lattice at higher temperatures, creep can become a significant mode of deformation at temperatures above $0.3\ T_m$.

2.7 THE EFFECTS OF ALLOYING

In the foregoing sections on the theoretical and actual strengths of metals, we have been talking about the crystal structures of pure materials, but the effects described also occur in alloys. An *alloy* is a mixture of two or more elements to form a crystalline structure that shows metallic properties. At least one of the components in the mixture must be a metal, but the other can be either metal or non-metal, provided that the bonding in the crystal is predominantly metallic in nature. Three possibilities exist:

    (i)   the species may have no effect on each other
    (ii) they may repel each other
    (iii) they may attract each other

Whichever occurs has a profound influence on the alloy's structure as well as its mechanical and other properties.

2.7.1 'Indifferent' atoms

If the atoms have no effect on each other they will be randomly dispersed within the crystal structure. There is a limit to the proportion of 'foreign' atoms that can be accommodated within a crystal lattice. As a general rule, if the difference in size between the species is 15% or less, extensive mixtures can be formed. In fact, if both elements in their pure state form the same type of crystal structure, complete intersolubility can occur. As the size difference increases beyond 15%, however, decreasing amounts of one element will be accommodated in the crystal lattice of the other.

The crystalline arrangement is called a *solid solution*. The predominant metal is called the *solvent* and the element accommodated is the *solute*. Most commercial alloys contain more than one alloying ingredient and the solid solution usually contains more than one solute.

The most common form of solid solution is the *substitutional solid*

solution, in which the solute atoms replace some of the crystal site:
normally occupied by the solvent. The less common form is the interstitiaɪ
solid solution in which the solute atoms fit into the spaces between those
of the solvent. This arrangement is called an *interstitial solid solution*.
Such solid solutions are formed only with small atoms, principally
hydrogen, carbon, oxygen, nitrogen and boron. Such interstitial solid
solutions formed between carbon and iron are the basis for the unique and
industrially extremely important alloys known as steels.

The metallurgical behaviour of interstitial solid solutions is quite
different from that of substitutional solid solutions. Substitutional solute
atoms merely distort the lattice and thereby impede the movement of
dislocations, giving a modest strengthening effect. Interstitial solute atoms
not only cause local lattice strains that obstruct the movement of
dislocations but can also cluster or form 'atmosphere' along dislocation
lines and pin them. Both of these effects strengthen the basic metal. If the
atmospheric forms, when the applied force reaches the elastic limit the
dislocations are suddenly torn away from the solute atoms. Such alloys
yield suddenly, a common and troublesome phenomenon in low-carbon
steels. Because interstitial atoms are small, they can diffuse easily through
the lattice and quickly form atmospheres again.

2.7.2 Atoms that repel

When the elements forming an alloy have a tendency to repel each other,
solid solutions are not favoured whatever the relative differences in their
atomic sizes. Instead, the atoms of each element form large aggregates of
their own kind that form distinct crystals within the bulk of the material.
These aggregates of atoms are usually large enough to be visible under an
optical microscope and form another constituent in the microstructure.

2.7.3 Atoms that attract

When the different atoms in an alloy attract each other, various
arrangements are possible. When the attraction is great, bonding between
the atoms may become ionic or covalent, leading to the formation of
chemical compounds that have no metallic properties. In such cases, nearly
all of the minor element combines with the appropriate portion of the
major element to form another constituent in the microstructure in the
shape of discrete particles within the crystal structure of the major
element. For example, if sulphur is present in iron that also contains
manganese, the sulphur will combine with the manganese (for which it
has the greater attraction), to form manganese sulphide. If there is
insufficient manganese present to take up all the sulphur, the remainder
will combine with iron to form iron sulphide. Discrete particles of
manganese sulphide and iron sulphide will be visible under an optical
microscope.

Chemicals that form in this way — oxides, phosphides, silicate, nitrides,
carbides, borides — are important hardening constituents in engineering

alloys and are stable because of the strong attraction between the atoms of which they are formed. They strengthen and harden alloys because, as with the introduction of large atoms into a crystal lattice described earlier, they distort the crystal lattice and pin the dislocations within it.

When different atoms in an alloy attract each other to a lesser degree, metallic bonding may still occur, but there may also be an effect called *ordering*. There is a strong tendency for unlike atoms in alloys of this type to arrange themselves in a regular 'ordered' pattern within the crystal lattice in such a way that each solute atom is surrounded by solvent atoms to which it is attracted. The basic crystal structure remains that of the solvent element. Ordering may extend over only a few lattice spacings — *short-range ordering* — or over the whole crystal lattice — *long-range ordering*. Ordered structures tend to be stronger and less ductile than random structures because of the distortion they cause in the crystal lattice requiring greater external forces to move dislocations through them.

The tendency to form ordered structures varies considerably with temperature. The strongest tendency occurs at temperatures well below the melting temperature and, as the melting temperature is approached, the ordered structure becomes random. The strengthening effect of an ordered structure can be used to advantage. Thus, certain alloys of gold and copper used in jewellery can be readily worked at a high temperature when their structure is random. When cooled, however, they assume an ordered structure of greater strength and hardness.

2.8 MICROSTRUCTURES

Having looked at the atomic and crystalline structure of metals, we now need to consider them in relation to their occurrence in bulk material. There is no distinctive stage at which large conglomerations of atoms begin to produce features that are visible under an optical microscope. However, before the advent of the electron microscope, metallurgists used the optical microscope to define the size limits of microstructural features. If a feature could not be detected using an optical microscope — in other words if it was less than 10^{-4}mm in size — it was in the atomic or molecular range. Anything larger, up to 1 mm in size, was termed *microstructural*, and features larger than 1 mm were termed *macro-structural*. This distinction remains in use today.

2.9 GRAIN STRUCTURE

Although it is possible to produce metals without a crystalline structure, those presently used in engineering, without exception, are crystalline. However, the regular crystalline structure is not normally continuous throughout even a small piece of metal, for there are discontinuities where one particular orientation of the structure changes to another. Each distinct volume of particular orientation is known as a *grain* and the

Fig. 2.10 Grain boundaries in etched section of a cast iron

disordered regions, between the grains, typically two or three atoms wide, are called *grain boundaries*.

Figure 2.10 shows a characteristic microstructure of a cast iron after the surface has been polished and then etched with chemicals that preferentially attack the grain boundaries. When examined under a microscope, the boundaries show up as lines and different constituents reflect the light in different ways, to give colour effects. In reality, the grains and boundaries are three-dimensional. The figure merely shows a section through them. The polyhedral bubbles formed by shaking a bottle of soap solution (or beer or milk!) are excellent three-dimensional models of grains and grain boundaries in metals.

Grains need not be regular polyhedra but may take various shapes depending on the thermal and mechanical history of the bulk material. The origin of their structure will be discussed later. The size of the grains has a major influence on mechanical properties. Other factors being equal, the smaller the size of the grains in a bulk metal the greater hardness, yield and tensile strength, ductility and toughness. The relationship between grain size and strength is given by the Petch equation:

$$\sigma_y = \sigma_i + k_y d^{\frac{1}{2}}$$

where σ_y is the yield strength, d is the grain diameter and σ_i and k_y are constants for the metal. k_y reflects the pinning effect of the grain boundaries and σ_i reflects the frictional force associated with drawing dislocations through the crystal. A graph of strength against grain size demonstrates the considerable increase in strength at fine grain sizes (Fig. 2.11).

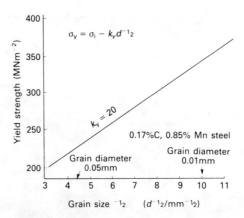

Fig.2.11 Relationship between grain size and yield strength for low carbon steel

2.10 DEVELOPMENT OF GRAIN STRUCTURE

The crystalline structure of a metal breaks down at its melting point. There are no grain boundaries and the liquid has no useful mechanical strength. The solid crystal structure re-forms on cooling the metal below its melting point. On solidifying, energy is given out in the form of latent heat of fusion, so the rate at which solidification proceeds depends on how quickly this heat of fusion can be removed.

The only technique known at present to produce metal without a crystal structure, i.e, amorphous, is to cool very small ribbons of liquid metal at rates exceeding 3×10^5 degs^{-1}. All other processes produce crystalline materials that exhibit some kind of grain structure.

When cooling is slow, a few of the large clusters of atoms in the liquid develop interfaces and become the nuclei for the solid grains that are to form. During solidification at a slow cooling rate, the first nuclei increase in size as more and more atoms transfer from the liquid state to the growing solid. Eventually, all the liquid has transformed and large grains have developed. The grain boundaries represent the meeting points of growth from the various nuclei initially formed. When cooling is rapid, many more clusters develop and each grows rapidly until it meets its neighbour. As a result more grains form and the grain size in the solid metal is finer.

The final grain size of a metal depends to a great extent, therefore, on the rate of cooling. When the rate of cooling cannot be controlled, in the central parts of large castings for example, a fine grain size can be secured by introducing nuclei into the liquid metal to cause crystal growth to begin at more numerous sites. This technique is known as *inoculation* or *grain refinement* and is of great importance for producing metal castings that are to be used in their cast condition, since grain size has so much influence on the mechanical properties.

The shape of the grains formed from the liquid metal is irregular and is influenced by three factors:

(i) Crystalline effects will, if sufficiently strong, dictate that the crystal develops fastest in certain directions and give rise to grains which may be columnar or angular, and needle-like in form.
(ii) Surface tension between the solid and liquid interface may dictate that the growing solid takes on a rounded external form which will ultimately give rise to polyhedral structures.
(iii) Thermal effects will generate rapid growth in grains with a large surface-area: volume ratio giving rise to spiky fir-tree-like growths known as *dendritic structures* (Fig. 2.12)

Unless there is deliberate introduction of foreign nuclei to promote a particular type of grain, cast metals usually exhibit a dendritic structure. The size of the dendritic grains and the spacing of the arms which form them depend on the cooling rate.

When a metal is strained beyond its elastic limit and is permanently deformed, part of the energy used during the deformation is stored within its grains in lattice distortions and dislocation tangles generated during the deformation. A cast structure formed directly from liquid contains no energy derived from mechanical deformation. Because of this, it will be stable and show little tendency to change after it has first formed. Even prolonged heating at high temperatures causes only a slight alternation in the shape of the grains. The exceptions to this rule (which leads to the popular misconception that heat treatment can alter the structure of castings) are iron and steel. In these metals, transformations of the solid structure can occur well below the melting point and they have the effect of refining the as-cast grain structure. However, in most engineering materials, no such transformations take place and the cast structure is retained until broken up by mechanical working.

2.11 EFFECTS OF MECHANICAL WORKING

Mechanical working of cast structures causes extensive deformation of the grains. It not only alters their physical shape but also introduces many dislocation tangles and part of the energy used to deform the metal is stored in these regions. If the working is carried out at high temperatures, atoms diffuse rapidly and dislocation tangles are annihilated almost as fast as they are formed. Consequently, there is no significant strain energy build up within the crystals. Instead, the cast structure breaks down and a new set of grain boundaries appear. The new grains tend to be polyhedral in shape and may be orientated in a direction of the working operation, as illustrated in Fig. 2.13(a). If the working is carried out at temperatures where diffusion is slow — usually at temperatures below about 0.4 T_m — dislocation tangles build up and some of the deformation energy is stored within the grains. The microstructure of broken-down crystals and strain

Fig. 2.12 Scanning electron microscope picture of solid dendrites growing in liquid metal, showing tree-like formations produced by spurs from a single initial nucleus. Growth has ceased in this example before the individual dendrites were able to meet and form grain boundaries, like those normally seen in planar sections. $250\times$

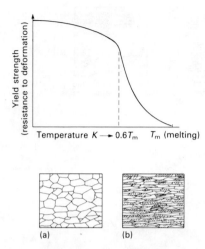

Fig. 2.13 Effect of temperature of deformation upon resulting grain structure. (a) Hot worked — above 0.6 T_m; (b) cold worked — below 0.4 T_m

bands is described as a cold-worked structure and is represented by Fig. 2.13(b). In this condition the strength is increased and ductility decreased in comparison with the initial state.

2.12 RECRYSTALLIZATION

The energy stored in a cold-worked structure makes it unstable. If it is heated to a temperature where diffusion becomes rapid, the dislocation tangles are eliminated and a new set of grain boundaries formed. The material is then said to be softened or annealed. The process is illustrated in Fig. 2.14. The nuclei for these new grains occur at sites within the previously broken down crystals where suitable blocks of the appropriate crystal structure persist. These blocks begin to grow into the deformed structure surrounding them until eventually the growing strain-free crystals meet. This process is called *recrystallization*. The greater the degree of stored energy — in other words, the greater the amount of cold-work the metal was subjected to — the greater the number of nuclei sites and the smaller the final grain size.

As indicated in Fig. 2.14 the properties of the annealed material now revert to those associated with the strain-free state, though strength and ductility are both improved compared with the cast condition, mainly because of the smaller grain size. (In practice, this recrystallization is not quite so simple. There is an intermediate stage not apparent under the optical microscope in which the dislocation tangles form low-angle grain boundaries. This intermediate stage is called *recovery* and although there is little change in mechanical properties from the as-worked state, there is a significant structural rearrangement at the atomic scale which precedes the

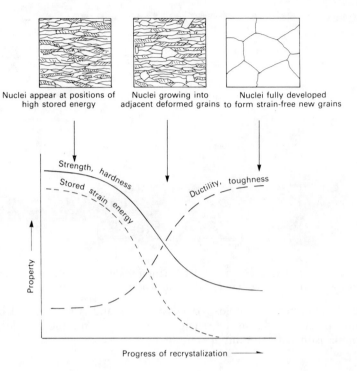

Fig. 2.14 Changes in microstructure and mechanical properties of deformed metal during recrystallization

microstructural changes described above.)

The *recrystallization temperature* is generally in the region of $0.4T_m$. The process of recrystallization is, however, time-dependent as well as temperature-dependent and in practice a temperature nearer to $0.6T_m$ is used so that the process proceeds rapidly.

Mechanical working can thus be seen to have a markedly different effect when carried out above or below the recrystallization range. If carried out below, the structure produced is distorted, contains stored energy, and is known as a *cold-worked* structure. If carried out above, the structure produced is softer, has mechanical properties similar to those in the annealed condition, and is referred to as a *hot-worked* structure. Note that 'hot' and 'cold' in these instances refers to the working temperature in relation to $0.6T_m$ or the recrystallization temperature, not to whether the deformation was carried out at or above room temperature. For example, lead is being hot-worked at room temperature whereas tungsten is being cold-worked if it is shaped at 1000°C.

Recrystallization of a cold-worked metal does not give a fully stable end-product. If heating is continued after recrystallization is complete the larger grains consume the smaller ones in order to reduce the overall grain boundary surface energy of the system. By cold-working the metal a relatively small amount (2–5%) to induce only a few high-strain energy regions to act as nuclei, and then annealing in the region of $0.8T_m$, it is possible to grow grains several centimetres in diameter. Although these may make very attractive exhibits, such large grain size should be avoided in all engineering products. The aim when annealing is just to complete the recrystallization process while retaining the smallest possible grain size, for the reasons evident from Fig. 2.11. Ductility is also greatly improved if the grain size can be kept small.

2.13 PHASE TRANSFORMATIONS

Casting, working, and annealing are not the only ways of developing microstructure in metals. Heat treatment alone can cause new crystals to form. Certain elements, such as iron, form different crystal structures at different temperatures. In such cases recrystallization occurs in order to accommodate the formation of a new crystal structure. For example, pure iron exists as a body-centred cubic space lattice (Fig. 2.3b) at temperatures up to 910°C, at which temperature the atoms rearrange themselves into a face-centred cubic space lattice (Fig. 2.3a). This is maintained up to 1390°C, at which point the crystals change back to a body-centred cubic arrangement. These atomic rearrangements bring about significant microstructural changes. They can be exploited to achieve refinement of grain size, but they can also cause grain-growth problems.

2.14 MICROSTRUCTURE OF ALLOYS

When atoms of different metallic elements are mixed together to form alloys an almost infinite variety of microstructures can be formed. Earlier, the possible interactions between solute and solvent atoms in forming solid solutions were shown. The purpose of this section is to extend the discussion to indicate the types of microstructure that may be encountered in industrial materials. Up until about 100 years ago knowledge of metals and alloys was very much a compendium of mixture compositions and processes, but metallurgists have since evolved a diagram known as a *phase diagram* that summarizes information about the structural variations that can rise when different metallic and non-metallic elements are mixed together. Phase diagrams are like maps that show the ranges of temperature and compositions over which the different crystal structures possible in the system may exist under equilibrium conditions. They are to the metallurgist what a drawing or blue print is to an engineer.

2.14.1 Phase diagrams

The various alloys of copper and zinc known as brass have long been important engineering materials, and there has grown up a multitude of descriptions for various compositions that possess different combinations of mechanical properties. For example, Red Brass, Gilding metal, Cartridge Brass, Clock Brass and Muntz Metal, are all mixtures of copper and zinc that may be represented on a phase diagram for the copper–zinc system up to 50% zinc. This 'metallurgical map' (Fig. 2.15) shows that, at room temperature, any one or a combination of any two or three crystalline states can be found in the copper-rich mixtures. When small amounts of zinc are added to copper, a solid solution is formed which is designated by the Greek symbol α, but this can exist only up to a composition containing 35% weight of zinc before further additions of zinc produce an additional and completely different crystal structure called β. The range of composition over which β-crystals form is from 35% to 59% by weight of zinc. Between 35% and 47% of zinc, a mixture of α- and β-crystals exists. Between 47% and 51%, β-crystals only are formed. A third type of crystal structure, called γ, occurs between 51% and 79% zinc and mixtures containing between 51% and 59% zinc consist of mixtures of β- and γ-crystals.

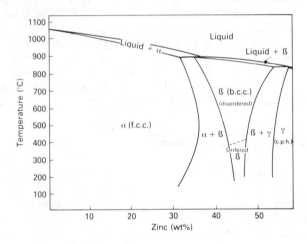

Fig. 2.15 The phase diagram for the copper–zinc system

The α-brasses, being solid solutions of zinc in the crystal lattice of copper are very ductile, soft and malleable, and are suitable for making thin sections, such as strip or wire. Increasing amounts of zinc raise the strength and make it possible to obtain a range of mechanical properties in brasses that are all easily formable and which, therefore, can be work-hardened. The β-phase brasses are ductile at high temperatures but the solid solutions become ordered below about 450°C and show considerably less ductility when cold than the α-brasses. Hence, β-brasses are suitable for applications where high tensile strength is required but must be hot-worked because of their restricted ductility below 450°C. The γ-phase is hard and brittle both at low and high temperatures and cannot be worked, though if small quantities of the γ phase are present in predominantly ß phase alloys, hot-working is possible.

The phase diagram tells us what proportions of crystal structures to expect but says nothing about their form or distribution, both of which affect the properties of the alloy. Many structural possibilities may occur. Consider a mixture of, say, 60% copper and 40% zinc. The phase diagram shows that this mixture should exist as roughly equal quantities of α- and β-crystals but their form and distribution can be any of the following:

  (i)   islands of β in a matrix of α
 (ii)   islands of α in a matrix of β
(iii)   equiaxed crystals of α and β
 (iv)   a coarse-grained cast structure containing long needles of α
  (v)   bands or strongly directional arrangements of groups of α- and β-crystals
 (vi)   α-crystals deformed by cold-working, or annealed and strain-free

Exactly which combination will occur in a particular piece of 60–40 brass depends on the thermal and mechanical history of the material. It is one of the skills of the metallurgist to be able to predict which structure will give the required engineering properties and show how that structure can be developed during, or be affected by, a manufacturing process.

The sloping line defining the region in the copper–zinc phase diagram in Fig. 2.15 labelled 'liquid' shows that the melting point of a copper–zinc alloy decreases with increase in zinc content. This change in melting point is not only of significance to the foundryman, but also to those who work and heat-treat the metal. They must work within the region delineated by the boundaries of the solid phases. Notice that the phase boundaries are not constant. They change with temperature and composition. Thus, 60Cu–40Zn brass consists of roughly equal proportions of α- and β-crystals at temperatures below 400°C but, because of the changes in solubility as the temperature is increased, the structure at 800°C would consist almost entirely of β-crystals with only a very small proportion of the α-phase. On cooling, the original roughly equal proportions should be restored, and this does occur provided that the cooling rate is slow enough to allow diffusing atoms to reorganize the crystal structure. If cooling is too fast, however, the initial proportions may persist. Although thermodynamically unstable, this structure can persist virtually indefinitely since diffusion rates below $0.3T_m$ are so low as to be insignificant. This has a fundamental and far-reaching influence on the potential to alter mechanical properties with heat treatments applied while the metal remains completely solid.

The phase diagram of the copper–zinc system has been used here to illustrate the importance and use of 'metallurgical maps'. Phase diagrams have been determined for most binary and more complex alloy systems and they are of great importance to the materials engineer when selecting an alloy composition and a production route for a given component.

2.15 STRUCTURAL CHANGES INDUCED BY HEAT-TREATMENT

The sluggishness of structural changes below $0.3T_m$ enables the metallurgist to maintain a remarkable degree of control over mechanical properties by the selection of appropriate heat-treatment. Two very important examples will be used to illustrate this, one based on aluminium alloys which can be *precipitation hardened* and the other on steels.

2.15.1 Precipitation hardening

Precipitation hardening is applicable to many alloy systems, but is most commonly applied to that of aluminium and copper, for which the phase diagram at the aluminium-rich end is shown in Fig. 2.16. Copper atoms dissolve in the crystal lattice based on aluminium, but there is a very significant change in solubility with temperature. At temperatures in the region of 500°C, nearly 5% by weight of copper can dissolve in κ in solid solution whereas below 300°C the amount which can be held in solution is less than 0.1%. When the solubility limit of copper in aluminium is exceeded, the excess copper forms a compound with aluminium designated θ and having the formula $CuAl_2$ and a totally different space lattice.

Thus with an alloy containing 4% by weight of copper held at 500°C long enough for all the copper to be taken into solution and then cooled rapidly by quenching in water to room temperature, there is no time for diffusion and rearrangement of the copper atoms as $CuAl_2$. All the copper in the aluminium is retained in solid solution. This condition will not change over a short period at room temperature, despite the fact that it is unstable. In this state is is known as a *supersaturated solid solution*.

The quenched material is in a so-called *metastable condition*. Its structure is that of a solution normally stable only at a higher temperature.

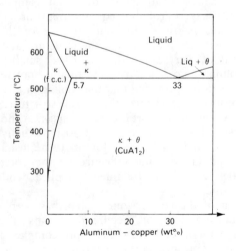

Fig. 2.16 Phase diagram for aluminium-rich end of the aluminium–copper system

If the temperature is now raised, slightly but to a level well below that at which all the copper will go back into solution, somewhere in the region of, say, 175°C, the excess copper atoms diffuse and cluster together. These clusters of copper atoms form regions of intense strain in the crystal lattice of the K aluminium–copper solution and obstruct the movement of dislocations, which results in an increase in hardness and strength. If the diffusion is allowed to continue discrete particles of compound CuAl₂ form, thus reducing the lattice strain and strength and hardness fall again. This sequence of events is shown in Fig. 2.17.

Precipitation hardening treatments are carried out in two stages, therefore:

(i) A solution heat treatment in which the material is heated to form a homogeneous solution whose composition is retained by quenching.

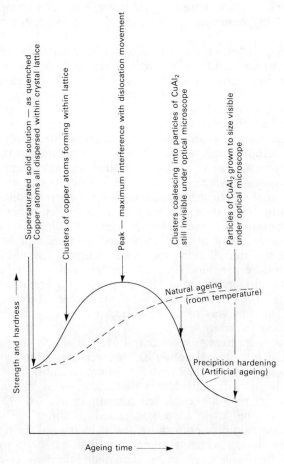

Fig. 2.17 The precipitation-hardening process

If the quench is not sufficiently rapid to prevent nucleation, or if all of the solute is not dissolved during the initial heating, the precipitation hardening response will be greatly diminished.

(ii) The metastable solution is then *aged* or *precipitation treated* at a temperature high enough to permit rapid interatomic diffusion and for a period of time that allows optimum clustering of solute atoms within the solvent crystal lattice. The critical size and dispersion of the clusters to produce the required degree of hardening will vary from one alloy to another.

The melting points of aluminium alloys are in the region of 600°C (873K) and room temperature is approximately $0.3T_m$, so some interatomic diffusion is possible even at room temperature, though only over a few lattice spacings and in time periods measured in days. Precipitation hardening will slowly occur at ambient temperatures, when it is known as *natural ageing*. This can raise the yield strength of a 4% copper alloy from about 60 MNm^{-2} immediately after quenching to 250 MNm^{-2} after five or six days.

Fortunately, nucleation and growth of $CuAl_2$ cannot take place at room temperature so the increase in strength is sufficiently permanent to be used in the design of engineering components. However, the structure is still highly unstable and, given an opportunity, the nucleation and growth process will continue. Hence, precipitation-hardened aluminium alloys that are to be subjected to moderate temperatures, as in the skin of supersonic aircraft, or repeatedly to high levels of stress, may *overage* in service and become much weaker than allowed for in the original design. This is a serious limitation on the use of high-strength aluminium alloys.

In alloy systems where $0.3–0.4T_m$ is *above* room temperature the agglomeration of solute atoms required to strengthen the alloys has to be brought about by heating the material to a higher temperature for a controlled period of time. This process is called *artificial ageing*. Aluminium alloys may be artificially aged to speed up, or give better control over, the precipitation process. For example, instead of five days at room temperature the same effect can be achieved in aluminium–copper alloys by 18 hours' treatment at 160° or by six hours' treatment at 180°C. In both cases, if the heat treatment is continued for too long overageing will occur. The only way to rescue overaged material is to subject it to a second solution treatment.

Artificial ageing offers greater control over an alloy's ultimate structure and properties but natural ageing has one very significant advantage. As strength rises during precipitation hardening, ductility falls. When the strength is highest the ductility is quite low — of the order of 5% elongation to fracture in a tensile test compared with 30%–40% when the material is in the 'as-solution-heat-treated condition'. The five days necessary to develop the optimum hardening by natural ageing means that there is time during which a component that has been solution-heat-treated can be formed, as when aluminium alloy rivets are used in assembling aircraft structures. If such rivets were fully precipitation-hardened by artificial ageing, they would be too low in ductility to undergo the rivetting operation, so they would crack.

In discussing precipitation hardening it has again been necessary to select as an example a material familiar to most engineers and widely used in aircraft structures. The technique is used extensively in many other alloy systems to provide the strength engineers require, and precipitation-hardened alloys are used in the most demanding applications in gas turbines, atomic energy plants, and similar critical locations, particularly when a combination of strength and properties such as corrosion resistance is required. The principles are the same but the temperatures are different for other alloy systems.

2.16 PHASE TRANSFORMATION IN CAST IRON AND STEEL

As we already know, iron is unusual among engineering metals because it can exist in the solid state in two crystalline forms. At low temperatures it takes up a body-centred cubic crystal structure. At 910°C, it changes to a face-centred cubic form, and then, just below its melting point, it reverts to the body-centred cubic form. The low temperature form is known as α-*iron* or *ferrite* and that which forms about 910°C is called γ-*iron* or austenite. Carbon, too is uncommon among the alloying elements, in that it is a very small atom that forms interstitial solid solutions with iron; in other words, it tucks itself into the spaces between iron atoms instead of replacing them in their space lattice. The maximum solubility of carbon in α-iron by weight is 0.05%, whereas the solubility in γ-iron is 2.0%. The phase diagram for the iron-rich end of the iron–carbon system is shown in Fig. 2.18.

Although the maximum solubility of carbon in *solid* solutions in 2.0%, carbon is even more soluble in liquid iron. The range of compositions from 2% to 4.5% carbon gives rise to the very important group of engineering materials called *cast irons*.

2.16.1 Cast iron

Cast irons have comparatively low melting points and are very fluid in the molten state. For these reasons, such alloys can be cast directly into moulds to form complicated shapes which require little machining to make them suitable for use in engineering structures. However, because of the variety of structures it is possible to produce during solidification and heat treatments, a remarkably wide range of property combinations is possible.

As cast irons solidify, any carbon in excess of 2% comes out of solution as graphite and a compound of iron and carbon of composition Fe_3C known as *cementite*. The brittle, needle-like crystals of cementite, as well as the flakes of graphite, weaken the structure in which they are held. For this reason, cast irons are usually brittle. However, with appropriate alloying agents, such as cerium or magnesium, the graphite can be made to precipitate in a spheroidal form that affects the strength and ductility to

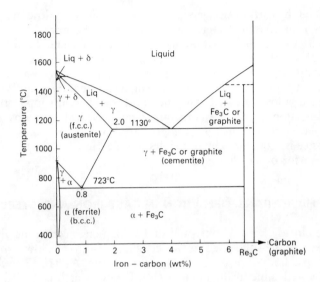

Fig. 2.18 Phase diagram of the iron-rich end of the iron–carbon system. Note that carbon can separate as graphite or as a carbide of iron, Fe$_3$C, depending on the rate of cooling

only a minor degree. If required, the formation of graphite can be suppressed in favour of cementite, making the material harder and more wear-resistant. Cast irons can be even further modified by heat-treatment to produce castings which are as strong and tough as many steels (i.e., the *malleable cast irons*).

There are so many variations of composition and treatment that a full description of cast irons is beyond the scope of this text, but they have been and will continue to be valuable engineering materials.

2.16.2 Steels

Let us turn our attention now to iron–carbon alloys that contain less than 2% carbon, alloys that form the basis of another group of materials of fundamental engineering importance–the *steels*.

If we examine the boundary of the γ-solid solution in the phase diagram for iron–carbon alloys in Fig. 2.18, we can see that the boundary falls from 910°C for pure iron to 723°C for an alloy containing 0.8% carbon, and then rises to 1130°C for an alloy containing 2% carbon.

What this means is that when an iron–carbon alloy containing more than 0.8% carbon cools to the γ boundary line, Fe$_3$C precipitates at such a rate that at 723°C, just 0.8% carbon remains in solid solution in γ iron (austenite). When an iron–carbon alloy containing *less* than 0.8% carbon cools to the γ iron boundary line, α-iron (ferrite) begins to separate first. Carbon is much less soluble in ferrite, so the carbon will build up in the remaining austenite at such a rate that at 723°C, the 0.8% solution of carbon in austenite is again achieved.

At 723°C, austenite containing 0.8% carbon always decomposes under equilibrium conditions in the same way to produce an intimate mixture of ferrite and cementite known as *pearlite*, on account of its irridescent appearance under the optical microscope (Fig. 2.19). The amount of pearlite increases from zero at very low carbon contents up to 100% if the steel contained 0.8% carbon to start with. Thereafter, it falls as the carbon content goes up to 2% because the excess carbon forms Fe_3C alone. Up to 0.8% carbon, the hardness and strength rise and ductility fails in almost direct proportion to the amount of pearlite.

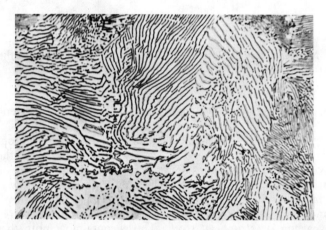

Fig. 2.19 Microstructure of pearlite — the intimate mechanical mixture of ferrite and Fe_3C formed at 0.8% carbon on slow cooling from austenite: magnification, 300×

2.16.3 The effects of heat treatment on steels

All the structures looked at so far relate to conditions in which sufficient time has been allowed for the carbon to diffuse and form the appropriate equilibrium constituents. Although strengths can be approximately doubled over the full range of carbon contents in these so called *equilibrium* or *stable* conditions, dramatic changes in properties can be brought about by treatments that suppress the attainment of, or allow a controlled approach towards, equilibrium. If, for example, a γ solid solution is quenched in water to prevent diffusion of carbon atoms, the carbon remains fixed in (though not part of) a lattice structure, setting up intense local lattice strains that block movement of dislocations. As a result, the structure is hard and extremely strong, but also very brittle. Under an optical microscope it looks like an array of needles, completely different from pearlite (Fig. 2.20). The needle-like structure formed when carbon is trapped by the iron crystal lattice is called *martensite*. It

Fig. 2.20 Microstructure of martensite formed by quenching steel from the austenite region: very hard and brittle: magnification, 300×

represents the maximum hardness obtainable with any given carbon content.

The degree of hardness of a quenched austenitic structure is proportional to the lattice strain. The lower the carbon content, the lower the strain. The maximum hardness appears to be reached in the region of 0.6 to 0.8% carbon. Industrially, steels in which the carbon is too low to give a useful hardening effect on quenching (below 0.25%) are known as *mild steels*, and those that do show a useful hardening effect are termed medium carbon or *structural steels*. Those containing more than 0.8% carbon not only offer greater strength and hardness on heat-treatment, but also contain excess cementite. Cementite particles confer excellent resistance to wear and these so-called *high-carbon steels* are often used for cutting and forming tools.

Unfortunately, the high hardness and strength of a wholly martensitic structure is difficult to exploit in practice because it is so brittle. A steel in this condition requires considerable support from a tougher material to be of use in service. An added disadvantage of quenched steels is dimensional change during transformation, due to the different lattice spacing of iron and carbon atoms in α- and γ-iron. Martensite expands by a few percent on formation, the expansion increasing with the carbon content. This expansion causes distortion and internal stresses that can lead to cracking. A water- or oil-quenched steel with more than about 0.4% carbon will almost certainly develop quench cracks, and the main reason why the welding of such steels is difficult is that cracking can occur even on cooling in air.

In certain situations this locked-in stress resulting from such transformations may be of put to good use, for, if an expansion takes place at the surface, the stresses set up are compressive and may considerably extend the fatigue life of, say, a gear tooth. Conversely, in regions where the locked-in stresses are tensile near or on the surface, a component could be on the brink of failure and require only a slight external load to cause sudden catastrophic cracking. The distortion caused by the locked-in stresses may be a considerable problem with precision components whose shape cannot be corrected by subsequent plastic deformation or machining because the material is too brittle. Consequently, no engineering component is used in a wholly quenched condition though it is commonplace locally to heat-treat components to produce the martensite structure in the surface layers and expect that the internal stress in the martensite will be compressive.

There are two ways of overcoming the brittleness of a quenched martensite steel to give desired combinations of hardness and toughness throughout the section. The first is to produce martensite and then *temper* it. Tempering is a controlled heat-treatment allowing some of the trapped carbon to escape from the interstitial spaces between the iron atoms and eventually form particles of cementite. The second is to cool the γ-iron from the high-temperature austenitic state in such a way as to cause it to change into another type of structure intermediate between the equilibrium pearlite and the metastable martensite. This intermediate structure has properties resembling those of a tempered martensite and is called *bainite*. In bainite some of the carbon atoms remain in the lattice, trapped between the iron atoms, and some are precipitated as a compound with iron. The particles of this compound are so fine that they become visible only with the help of an electron microscope. Under an optical microscope the structure, when suitably etched, appears as a dark mass of needles or acicular blocks (Fig. 2.21).

The temperature at which tempering is carried out is critical. Between 200°C and 300°C diffusion rates are slow and only a small amount of

Fig. 2.21 Microstructure of bainite transformation product of austenite: fairly hard and reasonably tough: magnification, 300×

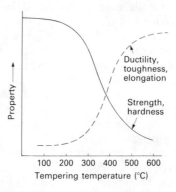

Fig. 2.22 Property changes with tempering of a martensitic (quenched) carbon steel. Notice how the most drastic changes take place between 300 and 400°C

carbon is released. As a result the structure retains much of its hardness but loses some of its brittleness. Between 500°C and 600°C diffusion is much faster, but allows most of the carbon atoms to diffuse from between the iron atoms to form cementite. The dramatic change in mechanical properties brought about by tempering a martensitic 0.4% carbon steel is shown in Fig. 2.22.

2.16.4 The effects of alloying additions in steel

All the structural changes described so far are brought about solely by the presence of carbon. But most steels used in engineering contain other elements, each added for a specific purpose. There are three main functions of alloying elements:

(i) To substitute for iron atoms in solid solutions, or in cementite, to increase strength, hardness and toughness. Alloying elements may also restrict the growth of the crystals or grains during transformation or heat-treatment. Certain elements are added to combine with impurities in the iron, such as sulphur or nitrogen.

(ii) To ensure the formation of martensite at lower cooling rates than those that operate when quenching in water. Heat can only travel from the centre of a piece of cooling metal to the surface at a limited rate, and if a section is above a certain thickness, the rate of cooling at the centre will be too slow for martensite to form. Small amounts (less than 5%) of chromium, nickel, molybdenum and vanadium, especially when used in combination, promote the formation of martensite so that thick sections can be cooled even in air and still form a martensitic structure in their centres. Alloying elements that exercise this function are said to increase the *hardenability* of the steel. Note that hardenability refers to a structural effect, not to the hardness level, which is dictated almost entirely by the carbon content.

(iii) To form alloy carbides that are harder and more resistant to wear than cementite (Fe_3C) and, in addition, to check the tempering of martensite. The steep drop in hardness of unalloyed carbon steels occurs between 300°C and 400°C (Fig. 2.22), but in steels containing tungsten, chromium, cobalt and vanadium this drop does not occur until the temperature reaches about 650°C. The difference means that such steels can be used in high-speed machining processes in which the tool may locally become red hot. Appropriate compositions of such steels are known as *high-speed tool steels*. Even with less highly alloyed materials, the functions of small amounts of additional elements in improving the toughness of tempered martensite is very important in engineering applications.

2.16.5 The effects of thermal treatments on low-carbon steels

As we already know low-carbon steels cannot usefully be strengthened by martensitic heat-treatment, but nevertheless they can be subjected to heat-treatment that refines their grain structure or relieves the effects of work-hardening. Low-carbon steels can be deformed and cold-worked in the same way as any other solid solution, but when further deformation would cause cracking, the material must be annealed.

In low-carbon steels, the transformation from γ- to α- iron takes place beetween 723 and 910°C and annealing can be carried out above or below this temperature range. The temperature chosen will have a significant influence on the structure produced. Figure 2.23(a) shows the structure of a cold-worked low carbon steel. If annealing is carried out below 723°C, the deformed ferrite (α-iron) crystals recrystallize but leave any pearlite as long *stringers* (Fig. 2.23b), because the temperature is too low for this to transform to austenite (see Fig. 2.18). Although the material is soft in this condition it has quite pronounced directional properties, because of the form assumed by the pearlite. The process is known as *sub-critical annealing*.

If annealing is carried out above 910°C the ferrite and pearlite recombine to form a γ solid solution and a new set of grains appears. These will have practically no relationship to the original ones. On cooling, equiaxed ferrite and pearlite reform to produce a structure similar to that shown in Fig. 2.23(c). The grain size of the γ phase is small when first formed but grows rapidly with time, and temperature. The smaller the γ grains the finer the resultant transformed structure, with consequent benefits as far as strength and toughness are concerned.

The effect of the transformation can be both beneficial and detrimental to the properties of a steel. It is beneficial in the sense that no matter what the starting structure of the material the small γ grains that form when the γ phase first appears subsequently give the material the optimum combination of strength and toughness. If, however, the austenite grains are allowed to grow large, ductility and toughness in the product fall

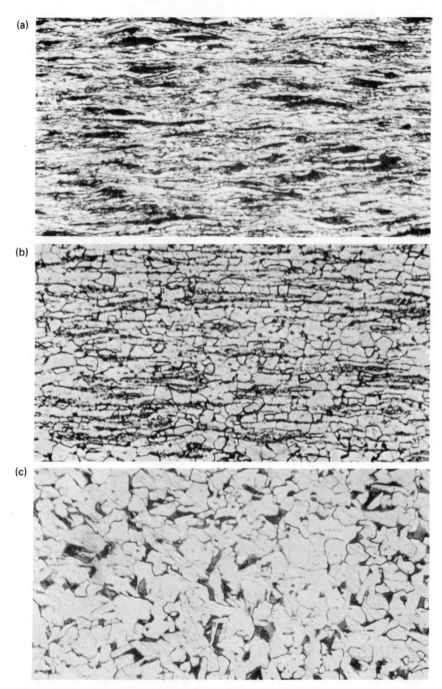

Fig. 2.23 The effect of annealing below and above the critical temperature for a cold worked low carbon steel: (a) as cold worked before annealing; (b) sub critically annealed at 675°C (note pearlite stringers in recrystallized ferrite matrix); (c) fully annealed at 920°C (entire structure equiaxed): all magnifications 175×

appreciably. Consequently, care must be taken to minimize the time available for grain growth.

There are two variations of the annealing process used with steels:

Full annealing

Full annealing is the description of the recrystallization process carried out above the critical temperature range and is used for the treatment of large complicated shapes, which are heated to the γ region for several hours to allow recrystallization. Cooling has to be slow and is usually achieved by leaving the parts in the furnace. This treatment is necessary to avoid the distortion that may be encountered if different parts of the metal cool at different rates. In such cases, it is found useful to have present in the structure particles that restrict the growth of austenite grains. This may be achieved with the addition of aluminium to the molten steel, when aluminium oxide particles form and obstruct grain growth. The aluminium is also added to lock up any residual oxygen in the steel.

Normalizing

Normalizing is an annealing process in which the steel is heated to a temperature 30–50°C above the critical temperature and for long enough for γ to form and the temperature to equalize throughout the section. It is then cooled in still air. This technique produces a much finer grain structure than full annealing and as air cooling is used minimizes the time a furnace is occupied. Normalizing is, therefore widely used as a means of refining the grain structure of steels. It is suitable for both cold-worked and hot-worked structures where the desire is to reduce the grain size by taking the steel through the transformation range. Where sectional size variations exist, full annealing may avoid problems with distortion or residual stresses.

Stress-relieving heat-treatment

As with other engineering metals, steels may also be given heat-treatment at low temperature (below $0.4T_m$) to allow for the recovery and relief from elastic strains introduced by forming operations such as machining, bending or welding, but structural changes are on a submicroscopic scale and are not visible using an optical microscope. To all intents and purposes no microstructural change takes place, only the relaxation of locked-in stresses.

Localized hardening of steel

Two requirements must be satisfied to produce the hard, wear-resistant martensitic structure suitable for use for bearing surfaces, and similar applications. First, the carbon content must be sufficient to cause lattice strain on quenching and second, the steel must be heated to a temperature above that at which the transformation from α- to γ-iron takes place so that all the carbon can be dissolved in the austenitic solid solution.

The martensitic structure that has the desired hardness and wear-resistance is also brittle and if the whole of the cross-section of a component were to be heat-treated the component would itself be wholly brittle. However, the production of a martensitic structure can be restricted to the surface layers, where its properties are in fact required while leaving the underlying material in a tough, but softer, condition. There are two ways of achieving this so-called case-hardening.

In the first method, known as *case-carburizing*, carbon, and sometimes nitrogen, atoms are diffused into the surface of a low-carbon steel until the carbon level reaches a value sufficient to form a hard martensite to the required depth. This is achieved by immersing the steel in a bath of molten chemicals of special composition at a high temperature or by heating the metal in an atmosphere of special composition, again at high temperature. The whole component is then quenched, and finally tempered to relieve the stresses induced by quenching. The central portions contain very little carbon and are not hardened by the quench. The outer case acquires its full martensitic hardness, however, and develops the required wear-resistance. By the use of nitrogen-containing chemicals in the molten bath or a furnace atmosphere containing nitrogen it is possible to incorporate nitrogen atoms in the surface layers of the steel along with the carbon. This process is known as *carbo-nitriding*, and gives additional hardness.

In the second case-hardening process, called *induction hardening*, the steel must initially have enough carbon throughout the section to form martensite when suitably quenched, i.e., more than 0.5% carbon. Before treatment, the steel must have a pearlite or tempered martensite structure to impart toughness and strength in the core material. The surface layer to be hardened, which may be quite localized, is then rapidly heated by electrical induction or by a gas flame to a temperature above the transformation point, which converts the surface layer into an austenitic solid solution. When this austenitic zone reaches the required depth, the whole component is quenched, causing the martensitic transformation. This transformation will occur only in the surface layers and not in the core because the latter did not reach the austenitization temperature. Hence, the final structure is hard martensite covering the original strong pearlite or tempered martensite. Because the core material has a high-carbon content this structure is stronger than a low-carbon steel carburized using the first method. A typical example of local case hardening — the tooth on a bandsaw blade — is shown in Fig. 2.24. The springy, medium-carbon steel of which it is made is cut to shape, and the teeth subsequently hardened by induction hardening. If the martensitic structure extended into the backing strip the blade would not flex properly, and unless the teeth are martensitic they would wear out very quickly as the saw was used. In this example, the tooth has a Vickers hardness of 900, and the backing a hardness of 280.

One other advantage of case-hardening is the volume expansion that takes place when martensite forms from austenite. When only the surface layers are subjected to the transformation as described above, they develop high residual compressive stresses when quenched, balanced by tensile stresses in the backing material. The net effect is that the compressive stress remains in the surface and any externally applied loads must

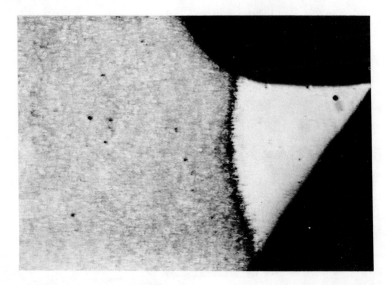

Fig. 2.24 Etched section of bandsaw blade showing locally hardened tooth on tempered steel blade: magnification 15×

overcome this residual stress before the surface layers themselves are subjected to tensile forces. Since fatigue cracking usually intiates at the surface under the action of tensile stress, fatigue life in particular is greatly improved by such treatment, and this is why numerous components used on motor vehicles, such as axle shafts and journals, are induction-hardened to depths of a millimetre or so. Not only do the martensitic surface layers improve the wear-resistance but the residual compressive stress greatly extends the life of the components.

2.17 MACROSTRUCTURE

The term *macrostructure* refers to features that are visible to the unaided eye, although in practice, magnifications of up to ten times may be used. The main purpose of macro-examination is to reveal features such as voids, inclusions, compositional segregation, fibrous structure, deformation and the effects of localised heat treatment, all of which can have a significant effect on engineering properties.

2.17.1 Voids

There are two common sources of voids: gas bubbles and shrinkage cavities. Bubbles of gas can get physically trapped as a metal is melted or cast, and gas can come out of solution as a metal cools and not escape

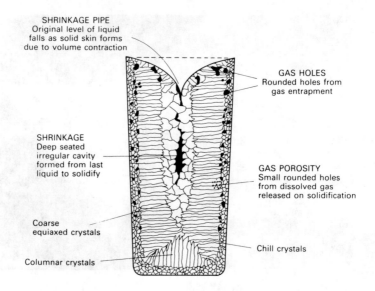

Fig. 2.25 Section of ingot depicting various macrostructural features

before solidification. In either case, smooth-walled cavities form in the solid metal. Shrinkage cavities arise because metals contract as they solidify. So, unless there is a constant supply of liquid metal to the molten centre of a casting, a cavity or cavities will be formed somewhere within the solid mass. The difference between the two kinds of voids is that shrinkage cavities have rough surfaces and tend to be intergranular.

These defects are shown clearly in Fig. 2.25, which illustrates the structure of a cross-section of a cast ingot. The outer layers of metal have cooled quickly and are formed of equiaxed crystals called *chilled grains*. Subsequently, the grains grew more slowly as the rate of cooling slowed down and assumed the dendritic or fir-tree pattern (Fig. 2.12) with the long axes of the grains pointing towards the centre line of the casting. As these dendritic grains formed, the liquid metal in the centre was cooling and nuclei were forming so that before the dendrites could meet at the centre of the casting, solidification occurred around the nuclei and a central zone of equiaxed crystals formed. Where the rate of cooling was sufficiently rapid, however (towards the bottom of the casting), the dendrites grew at a sufficient rate to meet in the centre and suppress the formation of equiaxed grains.

When the liquid metal first solidified at the walls of the mould the rate of cooling was fast and gas bubbles were entrapped, forming the rounded cavities shown in Fig. 2.25. Nearer the centre of the cross-section, irregular rough-sided shrinkage cavities can be seen. Near the top of the ingot the metal adjacent to the mould has solidified quickly but the molten central part has been drawn down to compensate for shrinkage in the lower part of the ingot. Here, a large cavity has formed known as a *pipe*. On occasions pipes may be formed low down in the ingot depending on how the cooling of the liquid metal progressed.

Figure 2.25 shows a schematic cross section of a rectangular shape, but similar defects are to be found in engineering castings. In the core of the ingot which is either hot- or cold-worked to make sheet, rod or other wrought-metal products, the voids may disappear during the working processes as the void surfaces weld together. If they do not, however, they form discontinuities in the wrought metal that weaken it seriously.

Voids in a shaped engineering casting effectively reduce the cross-section of the metal and weaken it. Further, if there are several connected voids present through the section of the casting, the casting will be porous to any gas or liquid it may be required to contain.

2.17.2 Inclusions

When a metal solidifies it may entrap non-metallic foreign matter. This may take the form of microscopic particles of oxide formed during the melting, globules of the molten-glass-like compounds used to protect the metal surface when melting, or even large pieces of ceramic that have broken away from the furnace roof or walls. These *inclusions*, as they are called, weaken the material by reducing its stress-bearing capacity or acting as notches from which cracks can develop.

2.17.3 Segregation

Segregation is defined as a gradual change in chemical composition across a section of metal and is normally encountered in castings. On the macroscale, compositional variations extend over several millimetres or even centimetres, and usually occur in one of three forms.

Gravity segregation

Gravity segregation is a separation of constituents during freezing caused by differences in density. One of the best known examples is the separation of cuboids of the compound SbSn in the lead–tin–antimony alloys used for making white-metal bearings, as illustrated in Fig. 2.26. These alloys solidify over a wide temperature range, during which cuboids of SbSn separate from the liquid. Since they have a much lower density, they float towards the top of the casting. However, the bearing properties depend on a uniform dispersion of these hard cuboids in the softer surrounding metal. The metallurgical solution is to introduce a small quantity of copper as an additional alloying ingredient. The copper causes another intermetallic compound to form, Cu_6Sn_5, which separates as star-like interlocking chains forming a fine mesh within the liquid. As a result, when the SbSn cuboids form the mesh prevents them from floating, and they remain evenly dispersed throughout the casting. Gravity segregation may also occur with immiscible liquids. For example, when lead is added to copper alloys to improve machinability, the denser immiscible lead may

Fig. 2.26 Section of small casting showing gravity segregation of cuboids of antimony–tin constituent which are of lower density than tin-rich matrix: magnification, $5\times$

not be uniformly dispersed throughout the liquid. Not only is the strength of the alloy affected but the improved machinability is not achieved.

Normal segregation

Normal segregation is the gradual increase in the proportion of low-melting-point constituents that accumulate in the last portions of a casting to solidify. The first solid crystals that solidify are rich in high-melting-point alloy constituents and, as solidification proceeds, the advancing solid front continuously pushes the low-melting-point ingredients into the remaining liquid.

Inverse segregation

Inverse segregation is the reverse of normal segregation inasmuch as the low-melting-point constituents appear on the *outside* of the casting. The explanation of this is that as solidification begins segregation develops normally, but, near the end of solidification, the liquid that has become enriched in low-melting-point constituents is forced between the dendritic grains to the outside of the casting. There are various reasons for this — an increase in pressure in the centre caused by a sudden release of gas in the remaining liquid metal, thermal contraction of the solid outer skin that builds up pressure on the remaining liquid, or capillary action in the spaces between contracting columnar crystals.

Low-melting-point constituents usually produce brittle intermetallic-type compounds so inverse segregation on the surface of the ingots which are to be hot- or cold-worked is a serious defect. If the segregation is not removed, cracking during subsequent working operations is likely. The problem is often troublesome with slabs of high-strength aluminium alloys.

In steels, minor segregation of alloying elements may give rise to 'banding' in which different response to heat treatment gives rise to differing structures (Fig. 2.27).

Fig. 2.27 Segregation of alloying elements in steel tool showing different structures produced by tempering response; This effect is called 'banding': magnification, 300×

2.17.4 Fibre structure

When metals containing voids, inclusions and segregations are hot- or cold-worked, these discontinuities and variations are drawn out in the direction of working and develop a fibrous structure rather like the grain in timber. They cannot be removed entirely by any subsequent treatment.

Toughness, tensile ductility, and fatigue resistance are considerably superior parallel to the grain compared with properties across the grain. For these reasons, great care needs to be taken when designing dies for forgings to ensure that the fibre structure flows with the shape of the component.

It is also highly undesirable to have outcrops of fibrous structure in regions of a component that are subjected to high service loads. The grain structure has to end somewhere, but the design must be such that it ends in an innocuous position. The sketches in Fig. 2.28 indicate a desirable and an undesirable fibre structure in a crankshaft.

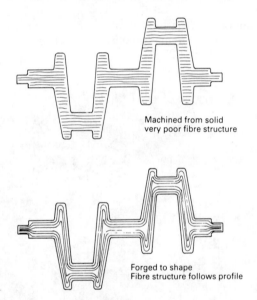

Machined from solid
very poor fibre structure

Forged to shape
Fibre structure follows profile

Fig. 2.28 The importance of correct fibre structure in a crankshaft. Note how machining cuts through flow lines whereas forging causes them to follow the contours

Fibre is an important structural feature in both large and small metal components and its development during working is one of the reasons why wrought products are preferred to castings, in which no directionality of fibre is possible. Although its importance in large items such as crankshafts and forged products is obvious, it is equally significant on a smaller scale. For example, screw threads can be made by machining wrought bar stock, in which case the fibres will run across the machined thread and end at the thread surfaces. However, threads can also be formed using a process known as *thread rolling*, in which the metal is forced to flow to make the thread contour. In this latter case the fibrous structure parallels the surface profiles of the threads and, made in this way, they exhibit significantly better toughness and fatigue resistance than machined threads.

2.18 MISCELLANEOUS EFFECTS OF HEATING

Carburization increases the carbon content of the surface layers of a steel but the reverse process, a reduction in carbon content, can occur when steels are heated in oxidizing atmospheres. Loss of carbon from the surface layers of steel can cause serious weakening of the surface region and the properties of the whole component may be adversely affected. Heating and cooling is inevitable when joining operations such as soldering, brazing and welding are performed.

2.18.1 Soldering

Soldering consists of joining metals with molten metal, later allowed to solidify, but is carried out at temperatures which cause no structural changes in the parent metals.

2.18.2 Brazing

Brazing is an operation conducted at temperatures that, although lower than the melting-point of the metals to be joined, are high enough to cause structural changes in some metals. The brazing metal melts and then solidifies to produce a 'cast' structure.

2.18.3 Welding

Welding consists of joining the parent materials by fusing them locally and frequently involves the introduction of filler metal to fill gaps between the parent materials. Consequently, welding always causes major structural changes in local regions of the materials being joined. These changes range from the production of cast structures in the fusion zone to heat-treated structures in the heat-affected zone which blend into the unaffected structure of the parent material. Figure 2.29 shows a typical etched cross-section through a weld in steel which shows all these distinct regions and in which successive runs have modified the cast structures of the earlier ones.

There are numberous types of welding processes that are carried out on a whole range of metallic materials, and they are far too diverse to be included here. With all of them, microstructural and macrostructural examination provides essential information relating to weld quality and the effects of the heating cycle on the joined materials. Virtually the entire range of structures that have been outlined in this chapter can be created

Fig. 2.29 Etched section of multi-run weld, showing successive weld pools and heat affected zones

in welded sections, since welding is basically a casting process carried out on a miniature scale with great variations in cooling rate. At the same time the surrounding solid material is heat-treated and so distortions and internal stresses are introduced by thermal contraction and expansion. Welding is a microcosm of the physical changes that may occur in the solidification and heat treatment of bulk materials.

SUGGESTIONS FOR FURTHER READING

W.O. Alexander and A.C. Street (1984) *Metals in the Service of Man*, Pelican, London.
A.H. Cottrell (1975) *An Introduction to Metallurgy*, Edward Arnold, London.
J.E. Gordon (1979) *The New Science of Strong Materials*, Pitman, London.
E.J. Rollason (1973) *Metallurgy for Engineers*, Edward Arnold, London.
R.E. Smallman (1970) *Modern Physical Metallurgy*, Butterworths, London.
H.A. Tyler (1980) *Science and Materials*, Van Nostrand, Wokingham.

CHAPTER 3

Properties of importance
in engineering design

In Chapter 1 engineering design was presented as an iterative procedure that involved a compromise between functional requirements, technical and economic constraints and performance.

Important considerations in the design process involve the mechanical, physical and chemical properties of the materials to be used. Some of the properties are structure-sensitive and others structure-insensitive. In either event the performance in-service may not be determined solely by the material's parameters identified and used in the initial design analysis, but may be influenced by the way in which the manufacturing processes modify these parameters. Thus it is necessary not only to give attention to the basic properties of interest in engineering design, but also to recognize the limitations imposed on these properties by the overall manufacturing and assembly process. In this chapter these factors will be considered in turn. First, attention will be given to the bulk and surface properties of materials. Subsequently the interaction of the stresses produced by service loading with the structures will be examined. The chapter concludes with consideration of the limitations to which materials are subjected by manufacturing conditions.

These considerations are complicated by isotropy and anisotropy of properties. It is conventional in engineering design to consider materials as comparatively homogeneous and isotropic in their bulk properties. That is, the properties of interest are not seen to vary with direction in the solid material. In practice some degree of anisotropy is commonplace, either because of microstructural variations or because of non-randomness in the distribution of grain orientations. It is often assumed that isotropy is most desirable. However, it is becoming increasingly clear that controlled anisotropy can frequently be very advantageous. After a description of the bulk properties the sources and consequences of anisotropy will be briefly examined.

3.1 STRUCTURE-SENSITIVE AND STRUCTURE-INSENSITIVE PROPERTIES

Before embarking on an examination of the properties of interest it must be made clear what is meant by structure-sensitivity and structure-insensitivity in so far as materials properties are concerned.

Structure-insensitive properties are those that are not influenced *significantly* by changes in microstructure or macrostructure. It is recognized that many of the physical properties of a material, e.g., bulk density, specific heat, coefficient of thermal expansion, do not vary other than by small amounts from specimen to specimen of a given material even if the different specimens have been subjected to very different working and/or heat-treatment processes. This is despite the fact that these processes may have produced quite substantial microstructural and macrostructural modifications. On the other hand most of the mechanical properties are very dependent on these modifications. Thus, for instance, the yield strength, ductility and fracture strength are seen to be structure-sensitive. Various aspects of this sensitivity were described in the previous chapter.

3.2 BULK PROPERTIES

3.2.1 Mechanical properties

There are several mechanical properties that determine the suitability of a given material for a given application. These can be either monotonic properties such as those describing the behaviour under simple tensile or compressive loading (including creep behaviour during deformation at elevated temperature) or cyclic properties associated with fluctuating loading under fatigue conditions. In all cases mechanical properties are most properly determined using standardized test pieces and standardized testing procedures. In considering the different properties it is conventional to distinguish between the time-independent behaviour normally encountered at ambient and intermediate temperatures and the time-dependent behaviour characteristic of creep deformation under load at temperatures above approximately half of the absolute melting temperature $(0.5T_m)$.

In identifying the mechanical properties of interest, attention must be given to the stress state to which the material is to be subjected. There are two simple stress states:

(i) tension or compression produced by uniaxial loading (Fig. 3.1a)
(ii) shear produced by torsional loading (Fig. 3.1b)

The two other common states of stress are biaxial stress such as that occurring in a vessel subjected to internal stress (Fig. 3.1c) and hydrostatic stress (Fig. 3.1d) resulting from immersion in a fluid. Although service conditions most usually involve complex combinations of these various stress states it is quite usual for one of the stress conditions to be so dominant that one particular mechanical property is of overriding importance.

When under stress, a material that is not stressed to fracture will deform either elastically or plastically. Elastic deformation is reversible on unloading while plastic deformation is permanent and is not removed on

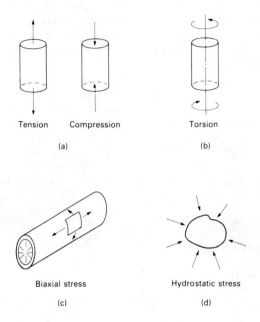

Fig. 3.1 Loading conditions leading to various stress states: (a) tension or compression; (b) torsion giving rise to shear stress; (c) internal pressure leading to biaxial tension; (d) hydrostatic stress

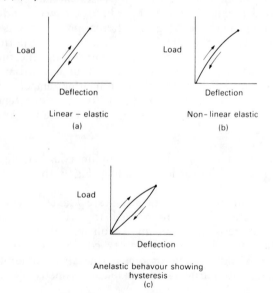

Fig. 3.2 Forms of elastic deformation exhibited by solids; (a) linear elastic behaviour; (b) non-linear elastic behaviour; (c) non-linear anelastic behaviour showing hysteresis

unloading. Frequently the elastic behaviour is linear and described as Hookean in that it obeys Hooke's law, namely that the deflection or displacement produced under load is directly proportional to that load. Non-linear elastic behaviour can also occur. Further, it is also possible for materials to behave anelastically in that the load–deflection path followed on loading is different to that followed on unloading and *hysteresis* occurs. In practice all metals are anelastic to some degree even in the apparently truly elastic regime. These various forms of elastic behaviour are illustrated in Fig. 3.2.

Tensile and compressive properties

Tensile properties are usually measured using shaped test pieces of circular or rectangular cross-section. The load is applied through threaded or shouldered ends (Fig. 3.3). The dimensions are commonly specified in detail. Typical test piece geometries are shown in Fig. 3.4. The important dimensions are:

(i) the *initial gauge length*, L_o, which is the prescribed part of the cylindrical or prismatic section of the test piece on which the elongation (the increase in length) is measured at any moment during the test
(ii) the *initial cross-sectional area*, S_o

These dimensions are not usually treated independently. Rather the initial gauge length is commonly taken to be $5.65 \sqrt{S_o}$ which, for a circular cross-section test piece, is a gauge length of 5 diameters.

In the most general case the elongation of the test piece comprises two principal contributions, an *elastic* and a *plastic* contribution. The elastic contribution contains both time-independent and time-dependent (anelastic) parts although the latter are normally neglected. The plastic

Fig. 3.3 Typical forms of tensile test specimen geometries: (a) threaded-end sample; (b) shouldered-end sample

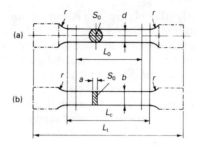

Fig. 3.4 Standard test piece geometries (after BS18: Part 1: 1970): (a) circular; (b) rectangular

contribution involves both a region of uniform deformation in which all parts of the gauge length elongate to the same amount and a non-uniform region in which localized deformation or *necking* occurs. These three main regions are indicated on the load–elongation curve shown schematically in Fig. 3.5, and a necked but unfractured test piece is shown in Fig 3.6. The deformation process is terminated by fracture in the necked region. Occasionally fracture will occur in the elastic region, and in such a case the metal under test would be described as being brittle. This is also the term normally applied to specimens that fracture after only a very limited amount of plastic deformation. When considerable plastic deformation occurs the metal is described as *ductile*. Even then, the final fracture can occur when little reduction has occurred in the necked region.

Load–elongation curves are normally converted to stress–strain curves taking into account the test piece dimensions. The various parameters of interest are defined as follows:

(i) the nominal or engineering stress, σ_n, is the load at any instant divided by the original cross-section area

(ii) the true stress, σ_t, is the load divided by the instantaneous cross-sectional area

(iii) the nominal or engineering strain, ε_n, is the ratio of the change in length

(iv) the true strain, ε_t, is the incremental instantaneous strain integrated over the whole of the elongation

Thus for a specimen of initial cross-sectional area, S_o, and initial gauge length, L_o, elongated uniformly under a load P to a gauge length L with a corresponding cross-sectional area S, the appropriate relationships are:

$$\sigma_n = \frac{P}{S_o} \qquad\qquad \sigma_t = \frac{P}{S}$$

$$\varepsilon_n = \frac{L\text{-}L_o}{L_o} \qquad\qquad \varepsilon_t = \int_{L_o}^{L} \frac{dl}{l} = \ln\frac{L}{L_o} \qquad\qquad (3.1)$$

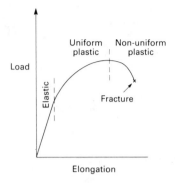

Fig. 3.5 A typical load–elongation curve (schematic) for tensile deformation

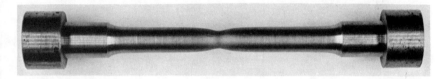

Fig. 3.6 A necked, un-fractured tensile test specimen

During uniform elongation the volume remains constant, i.e., $S_o L_o = SL$. Using this relation it can easily be shown that for tensile deformation:

$$\sigma_t = \sigma_n (1 + \varepsilon_n)$$

$$\varepsilon_t = \ln (1 + \varepsilon_n) \tag{3.2}$$

For many metals the transition from elastic to plastic deformation is not clearly evident but occurs progressively (as shown schematically in Fig. 3.5). Then the stress at which the metal is said to yield or flow plastically is defined in terms of a *proof stress*. The proof stress is the stress at which a permanent elongation (or permanent set) of a specified percentage of the initial gauge length (usually 0.2%) occurs. The method of determining the proof stress is given in Fig. 3.7.

Some metals, however, especially steels, show an abrupt yield point followed by a short period of non-uniform plastic strain. The stress–strain curve then appears as shown in Fig. 3.8. During the extension immediately after yielding a series of markings known as Lüder's lines or Lüder's bands are observed on the surface of the test piece. These bands indicate the regions which are deforming plastically and they broaden until they occupy the whole of the gauge length. This occurs over the Lüder's strain (normally about 5–6%) (see Fig. 3.8). When the whole of the gauge length has yielded the behaviour then is similar to that of metals which do not exhibit a yield point phenomenon.

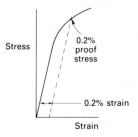

Fig. 3.7 Determination of the proof stress from a tensile stress–strain curve

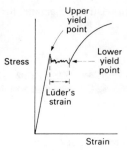

Fig. 3.8 Schematic stress–strain curve showing a sharp, upper yield point and a lower yield point. The Lüder's strain during which the plastic regions spread to occupy the whole specimen is shown

In a steel that shows a yield-point phenomenon two parameters are defined to describe yielding:

(i) the *upper yield stress*, which is the stress at the initiation of yielding
(ii) the *lower yield stress*, which is the lowest value of stress during propagation of the Lüder's bands, ignoring any initial transient effects that might occur.

When the transition from uniform to non-uniform plastic deformation occurs and necking begins the load-elongation curve goes through a maximum. This leads to the definition of a property, the ultimate tensile stress, σ_{uts}, given by:

$$\sigma_{uts} = \frac{\text{Maximum load}}{\text{Initial cross-sectional area}} = \frac{P_{max}}{S_0} \qquad (3.3)$$

Two other important mechanical property parameters determined from the tensile test are:

(i) the *percentage elongation to fracture*, which is the permanent elongation of the gauge length after fracture expressed as a percentage of the original gauge length

(ii) the *percentage reduction in area*, which is the ratio of the maximum change in cross-sectional area that has occurred during the test to the original cross-sectional area.

Then if L_u is the gauge length after fracture and S_u is the minimum cross-sectional area in the necked region:

$$\% \text{ elongation } = \frac{100(L_u - L_o)}{L_o}$$

$$\% \text{ reduction in area } = \frac{100(S_o - S_u)}{S_o} \tag{3.4}$$

Data for mechanical properties are available from various sources, including published standards and reference books (such as *Metals Handbook*, Vol. 1, or the *Metals Reference Book*). Some typical data for various common metals and alloys (excluding steels) are shown in Table 3.1.

Table 3.1 Typical mechanical property data for some common metals and alloys

Metal or alloy	0.2% proof stress (MNm^{-2})	Ultimate tensile strength (MNm^{-2})	% elongation to fracture ($L_o = 5.65 \sqrt{S_o}$)
Aluminium (annealed)	34.0	77.2	47
Aluminium (cold-worked)	94.2	115.8	13
Al–Mg–Si alloy (annealed)	54.1	123.5	25
Al–Mg–Si alloy (age-hardened)	262.5	308.9	12
Duralumin (annealed)	123.5	231.6	15
Duralumin (age-hardened)	278.0	432.4	15
Copper (annealed)	54.1	223.9	56
Copper (cold-worked)	285.7	316.6	13
70–30 brass (annealed)	84.9	319.7	65
70–30 brass (cold-worked)	378.4	463.3	20
Phosphor bronze (annealed)	123.5	339.7	66
Phosphor bronze (cold-worked)	640.9	710.4	6
Magnesium (annealed)	61.8	185.3	5
Nickel (annealed)	139.0	478.7	40
Titanium (annealed)	247.1	324.3	36

It is clear from Table 3.1 that the mechanical properties shown are structure-sensitive, being altered by treatments such as work-hardening or age-hardening. The mechanical properties of steels are also structure-sensitive, particularly for those alloy steels subject to controlled heat-treatments. Mechanical property data for various steels are given in Table 3.2

Table 3.2 Typical mechanical property data for steels

Steel	Condition	Yield stress (MNm^{-2})	Ultimate tensile strength (MNm^{-2})	% elongation to fracture ($L_o = 5.65 \sqrt{S_o}$)
0.15%C steel	Normalized	245	490	20
	Cold-worked	450	600	10
0.40% steel	Normalized	280	540	16
	Cold-worked	510	650	8
	Q and T*	465	850	16
0.55%C steel	Normalized	355	700	12
	Q and T	570	1000	12
1%Cr steel	Q and T	680	1000	13
1% Ni steel	Q and T	585	930	15
1¼%Ni–Cr steel	Q and T	755	1080	12
1%Cr–Mo steel	Q and T	850	1160	12
1½%Ni–Mo steel	Q and T	465	770	16
1¼%Ni–Cr–Mo steel	Q and T	1240	1540	5
2½%Ni–Cr–Mo steel	Q and T	1235	1540	7

*Q and T = quenched and tempered

Compressive mechanical property data are more difficult to determine than tensile data. This is because compressive testing is normally carried out on cylindrical specimens short enough to avoid buckling when an axial load is applied. Then, however, there are frictional end effects that influence the behaviour. There are various ways to minimize these end effects but none is completely satisfactory. For compressive loading, nominal stress and strain and true stress and strain are defined in similar ways to the corresponding definitions for tensile loading, with the exception that the strains must be arranged so as to produce positive values. Then:

$$\sigma_n^c = \frac{P}{S_o} \qquad \sigma_t^c = \frac{P}{S}$$

$$\varepsilon_n^c = \frac{h_0 - h}{h_0} \qquad \varepsilon_t^c = \ln\left[\frac{h_0}{h}\right] \qquad\qquad (3.5)$$

for compressive deformation of a cylinder with initial cross-sectional area, S_o, and initial height, h_o, deformed to a height h with a corresponding area, S. There is no region of non-uniform plastic deformation in compression to correspond to the region of necking in tensile deformation although the frictional end effects can give rise to 'barrelling' of the specimen at large strains. For that part of the deformation where uniform compression occurs constant volume can again be assumed, which leads to the relationships:

$$\sigma_t^c = \sigma_n^c (1 - \varepsilon_n^c)$$

$$\sigma_t^c = - \ln (1 - \varepsilon_n^c) \tag{3.6}$$

For a given material the nominal or engineering stress–strain curves in tension and compression and the true stress–strain curve (which is a material property) are related in the way shown in Fig. 3.9.

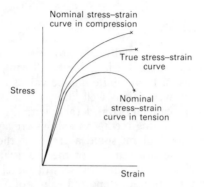

Fig. 3.9 Schematic nominal stress–strain curves in tension and compression with common true stress–true strain curve derived from them

Shear properties

The testing of specimens in order to determine mechanical properties in shear is usually carried out in torsion. For a solid cylindrical specimen there is a gradient of stress and strain across the cross-section increasing from zero at the centre to a maximum at the specimen surface. For this reason it is common to test thin-walled tubular specimens for which the stress and strain across the wall can be assumed uniform. Then, for a tube of radius r and wall thickness t subjected to a torque T the relationships for stress and strain are

shear stress $\qquad \tau = \dfrac{T}{2\pi rt}$

angular shear strain $\qquad \gamma = r\varnothing \tag{3.7}$

where $\varnothing$ is the angular twist per unit length along the tube axis. As with deformation in tension and compression during shearing there is a transition from elastic to plastic deformation at a stress known as the *shear yield stress*. Again the transition is usually gradual which requires the use of a construction similar to that used in the determination of the proof stress (Fig. 3.7) for the determination of the shear yield stress.

Elastic properties

The constant of proportionality between stress and strain in the linear elastic region during the initial stages of loading a specimen either in tension or compression is known as Young's modulus. This is usually given the symbol E and defined as:

$$E = \frac{\sigma}{\varepsilon}$$

(3.8)

For shear deformation the shear modulus is given by:

$$G = \frac{\tau}{\gamma}$$

(3.9)

These elastic moduli are normally taken to be structure insensitive for a given material. The variations attributable to alloying are small and thus on both counts it is usual to attribute unique values of the elastic moduli to a given material. However, it is well known that the elastic moduli of single crystals of metals can vary considerably with direction in the crystal. Since our common engineering metals and alloys are aggregates of crystals the bulk elastic moduli will be some average of the properties of the individual crystals. If the orientations of the grains in the aggregate are random then this will lead to singular values of the elastic moduli. These are the values normally reported. Reported values of Young's modulus for some common metals are given in Table 3.3. If the arrangement of crystals is non-random, then some of the anistropy of the single crystal elastic properties will appear in the polycrystalline material. This non-randomness is known as preferred orientation or *texture*, and can arise because of working and heat treatment processes. For example, in rolled and annealed sheet materials texture is the norm. The extent to which elastic anistropy is likely as a consequence of texture is, in cubic metals and alloys, determined by the factor $2C_{44}/(C_{11}-C_{12})$ where C_{11}, C_{12} and C_{44} are the single crystal elastic constants. Some values of this factor are given in Table 3.4.

Table 3.3 Typical values of Young's modulus for metals

Metal	Al	Au	Ti	Cu	Fe	Ni
E(GNm^{-2})	71	79	120	130	211	200

Table 3.4 Value of the elastic anisotropy factor for metals

Metal	Cu	Ni	Fe	Al	W	Mo	Nb
Elastic anisotropy factor	3.2	2.6	2.4	1.2	1.0	0.71	0.51

Possible variations in Young's modulus need to be borne in mind when the behaviour of metals and alloys that have been processed in such a way as to produce texture, is being examined. For instance, it has been shown that cold-rolled and annealed copper sheets can show variations in Young's modulus for specimens cut from the sheet ranging from 91 to 142 GNm^{-2} (E for copper is commonly taken to be 124 GNm^{-2}: see Table 3.3).

Hardness

Hardness is a measure of resistance to deformation under conditions in which a loaded indenter is forced to penetrate the surface of the metal under test. The deformation during indentation is again a combination of elastic and plastic behaviour. However, as measured the hardness is largely related to plastic properties and only to a secondary extent to elastic properties. Other measures of hardness are sometimes used, such as scratch hardness, which depends upon the ability of one solid to scratch or be scratched by another, or rebound (dynamic) hardness, which involves the dynamic deformation of a specimen expressed in terms of the part of the energy of impact absorbed when an indenter drops upon a specimen. However, indentation hardness is by far the most important of the measures in current use.

The different methods of indentation hardness testing differ in the form of indenter that is forced into the surface. The *Brinell* hardness test uses a hardened steel ball as the indenter. The *Vickers* hardness test is based upon the use of a square-based diamond pyramid of 136° included angle. The *Rockwell* hardness test involves a diamond cone indenter with 120° included angle and a slightly rounded point. In all cases the hardness number recorded is related to the ratio of the applied load to the surface area of the indentation formed.

The testing procedure involves bringing the indenter into contact with the surface of the material to be indented followed by a cycle of loading–holding under load–unloading. On removal of the load the size of the indentation is measured using a microscope. For the two most commonly used hardness measurements the hardnesses are then given by:

$$\text{Brinell Hardness Number (BHN)} = \frac{P}{\frac{1}{2}\pi D\ (D-\sqrt{D^2-d^2})} \qquad (3.10)$$

where P is the load in kg, D is the ball indenter diameter (mm) and d is the indent diameter (mm), and:

$$\text{Vickers Hardness Number (VHN)} = \frac{1.854\ P}{L^2} \qquad (3.11)$$

where again P is the load in kg and L is the average length of the diagonals of the pyramidal indentation. In practice Brinell and Vickers hardness machines utilise tabulated data that relate the direct measurement of d or L to the hardness according to the load (this usually lies in the range 5–30 kg).

Unlike Brinell and Vickers hardness measurements, Rockwell hardness measurements usually make use of a direct-reading machine.

For hardness measurements to be valid, they should not be sensitive to the conditions of testing. Normally there is a range of loads for which this requirement is met for a given range of hardness. For this reason it is common to impose restrictions on the permissible load ranges.

The different hardness measurements correlate quite closely, especially at lower values. Some typical values are given in Table 3.5.

Table 3.5 Typical values of hardness for some metals and alloys

Metal	BHN	VHN
Copper (annealed)	49	53
Brass (annealed)	65	70
0.40%C steel (normalized)	152	157
1½% Ni–Cr–Mo steel (Q and T)	380	400
Tool steel (Q and T)	670	720

The correlation of hardness values with other measures of resistance to deformation, such as tensile properties, is more difficult. For example, the deformation in a Vickers indentation is equivalent to a tensile strain of approximately 8%. At best the correlations are empirical, but even then they must be treated with caution, since the relationships derived assumed that the materials are of uniform composition and have been subjected to uniform heat and/or mechanical treatments. Where surface treatments have been carried out, e.g., surface-hardening by carburizing of nitriding, then no correlations are possible.

Despite the limitations of hardness measurements as absolute measures of mechanical properties, hardness testing can be used very effectively as a rapid means of materials assessment. This is particularly the case with quality control procedures or when qualitative evaluations are required.

Impact strength and fracture toughness

Often materials that show quite acceptable properties when tested in tension at slow loading rates fail in a brittle matter when subjected to rapid loading. Proneness to failure under conditions of fast fracture is enhanced if there are notches or other defects present on the surface of the sample. Ductile face-centred cubic materials, such as copper and aluminium, normally resist fast fracture under all loading conditions and at all temperatures. This is not the case, however, for many ferrous alloys, particularly plain carbon and low alloy steels. Of particular note is the occurrence of a ductile-to-brittle transition in these alloys as the temperature is lowered. Resistance to fast fracture is commonly referred to as *toughness*, and loss of toughness in service can have catastrophic effects.

It is important to be able to quantify toughness, and conventionally this has been done by carrying out high strain rate impact tests on notched samples with standardized specimens. The most common test is the Charpy test in which a 10mm square bar with a machined notch is struck by a calibrated pendulum (Fig. 3.10). The energy absorbed from the swinging pendulum during the deformation or fracture of the test specimen is used as a measure of the impact strength.

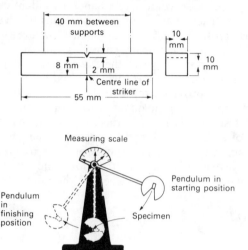

Fig. 3.10 A Charpy test specimen and the associated test procedure (after BS 131: Part 2: 1972)

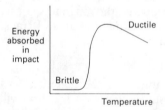

Fig. 3.11 The ductile-to-brittle impact transition curve for steel (schematic)

The nature of the resulting fracture surface can also provide information of value in assessing impact resistance. for instance, high impact strength is normally associated with a ductile, fibrous fracture, whereas low impact strength corresponds to a brittle, cleavage fracture.

The ductile-to brittle transition in steels can be best demonstrated by the change in the energy absorbed in fracture as a function of temperature. This is shown schematically in Fig. 3.11. The transition temperature is normally near room temperature, and this can have very important consequences in so far as service behaviour is concerned. Thus a structure or an engineering component can suffer sudden failure under service loading if temperature changes occur that move the operating conditions to below the transition temperature. The transition temperature is affected by both metallurgical variables, such as grain size (increasing the grain size raises the transition temperature and vice versa) and by service variables, e.g., the presence or absence of notches and/or residual stresses. The use of the notched impact test was adopted because this involved a combination of conditions most favourable to impact failure with low energy absorption. However, the impact energy is only a relative measure of impact strength. For many years there were efforts made to identify a quantitative measure of toughness which could be made use of by design engineers. This was provided through the advent of fracture mechanics.

The earliest approach to the quantification of fast fracture was provided as a result of tests on glass, an intrinsically brittle material. By assuming the existence of a sharp crack and balancing the energy released by crack propagation against the surface energy needed to create new fracture surfaces, the existence of a critical stress for rapid crack propagation was predicted. This critical stress was given by the relation:

$$\sigma_{crit} = \sqrt{\frac{E\gamma}{\pi a}}$$

(3.12)

where E is Young's modulus and γ is the energy per unit area of new surface. The assumed crack length is a for a surface crack or $2a$ for an internal crack.

Since the fracture of ductile or semi-brittle materials involves the dissipation of energy additional to that needed to create the new surfaces, Equation (3.12) needs to be modified for general use. This is done by defining a new materials property, the toughness G_c, which is a measure of the *total* energy absorbed in making unit area of crack. Equation (3.12) is then rewritten in the form:

$$\sigma \sqrt{(\pi a)} = \sqrt{(EG_c)}$$

(3.13)

in which the system variables and the materials variables are collected together. This equation states that fast fracture will occur in a given material subjected to a stress σ when a crack reaches a critical size, a, or, alternatively, it will occur in the presence of a crack of size a when a critical stress σ is reached.

The materials parameter $\sqrt{(EG_c)}$ which clearly plays a critical role in determining the likelihood of fast fracture occurring is usually given the symbol K_c and called *the fracture toughness*. The determination of the fracture toughness is also carried out using standardized test specimens and procedures. Values vary substantially from material to material and some representative values are given in Table 3.6.

The actual value of the fracture toughness to be used when assessing service behaviour is also dependent on the stress condition. Thus it is usually lower if failure occurs in plane strain rather than under a more relaxed combination of stresses.

Table 3.6 Fracture toughness values of different materials

Material	$K_c(MNm^{-3/2})$	Material	$K_c(MNm^{-3/2})$
Ductile metals, e.g., Cu	200	Medium C steel	50
Mild steel (room temp.)	140	Cast iron	12
Mild steel (-100°C)	10	Glass reinforced plastic	40
Pressure vessel steel	170	High density poly	2
Titanium alloys	70	Reinforced cement	12
		Glass	0.8

Fatigue properties

When discussing the tensile and compressive properties of metals and alloys, it was assumed that the loading was monotonic, i.e., it increased steadily until failure occurred. In most practical situations this type of loading does not occur. Instead the applied stresses fluctuate, often in a random way. In some cases, such as with a rotating shaft subjected to bending stresses, the stress at a given point on the surface varies cyclically from tension to compression and back again. Various forms of cyclic stress loading are shown schematically in Fig. 3.12. In the presence of cyclic stress, it is frequently found that premature failure occurs at stress levels much lower than those required for failure under a steady applied stress. This phenomenon of premature failure under fluctuating stress is known as *fatigue*, and involves processes of slow crack growth.

Fatigue failure is a statistical process and as such its measurement requires extensive testing. Usually, and most simply, this is achieved by

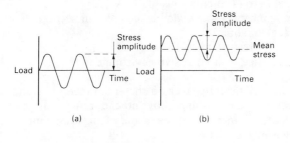

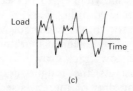

Fig. 3.12 Forms of cycling loading: (a) fluctuating tension-compression (mean stress = 0); (b) fluctuating tension (mean stress > 0); (c) random tension-compression loading

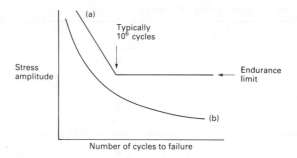

Fig. 3.13 S–N curve for zero mean stress loading for (a) steel, and (b) aluminium

preparing a substantial number of test specimens and subjecting them to defined cyclic conditions and measuring the number of cycles required for failure at a given stress. The cyclic loading conditions in most common use involve either fluctuating tension (tension–zero–tension) or push–pull (compression–tension–compression) stresses. The mean stress, σ_m, and the fluctuating stres amplitude, σ_a, are then the important test parameters. The test data are plotted as an S–N curve as shown in Fig. 3.13 for zero mean stress loading. With ferrous alloys a limiting stress below which failure does not occur is recorded. This is known as the *endurance limit*. However, this must be treated with caution since the fatigue behaviour is very sensitive to operating conditions. Thus:

   (i)   random stress fluctuations
  (ii)   stress concentrations, such as changes in section
 (iii)   surface roughness
 (iv)   residual stress
  (v)   the presence of a corrosive environment

can all have a deleterious effect on the endurance limit and fatigue life.

With non-ferrous metals and alloys no endurance limit is recorded, there being only a steady increase in life as the cyclic stress amplitude falls. This is also shown in Fig. 3.13.

Fractures caused by fatigue usually have a very characteristic appearance. A typical fracture surface is shown in Fig. 3.14 (see also Section 4.2.2). The point of crack initiation (arrowed in Fig. 3.14) is surrounded by a flat region over which slow crack growth has occurred. Macroscopically this region usually exhibits a series of well-defined markings known as 'beach marks'. On a microscopic scale slow crack growth is accompanied by the formation of fatigue striations. As the fatigue crack propagates, the effective stress on the remaining part of the cross-section increases until there is sudden fast fracture giving rise to a typically fibrous ductile fracture. This is also apparent in Fig. 3.14. It is now common practice to describe fatigue crack propagation after crack initiation by relating the

Fig. 3.14 A typical fatigue fracture surface showing clearly defined 'beach marks' and a region of rapid, ductile fracture

rate of crack growth per cycle (da/dN) to the cyclic stress intensity, $\Delta K (= K_{max} - K_{min})$ through an equation of the form:

$$\frac{da}{dN} = A\Delta K^m$$

(3.14)

where A and m are material constants. Equation (3.14) applies to the steady-state part of the growth regime illustrated in Fig. 3.15.

Apart from the factors described above, the mean stress has a significant effect on the fatigue life. Thus when, for a given cyclic stress loading pattern, the mean stress, σ_m, and the stress amplitude, σ_a, are defined (see Fig. 3.12) the endurance limit can be determined using the Goodman relationship or some equivalent relationship, such as that defined by Gerber. These are shown in Fig. 3.16.

All of these considerations do not make allowance for the influence of metallurgical factors. It is well established that microstructural variables

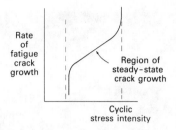

Fig. 3.15 Schematic relationship between rate of crack growth per cycle and the cyclic stress intensity under fatigue loading conditions

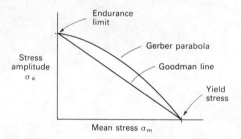

Fig. 3.16 The relationship between permissible fatigue stress amplitude and the mean stress according to Goodman and Gerber

such as changes in grain size, alloying and most particularly, the presence or absence of non-metallic inclusions, also have significant effects on fatigue life.

Creep and stress-rupture properties

As the temperature of measurement of plastic properties is increased it is observed that properties, such as the yield strength and the ultimate tensile strength fall. At temperatures above about $0.5 T_m$ the behaviour becomes time-dependent and both metals and alloys are found to elongate under constant stress. This is known as creep deformation, and a typical family of creep curves is shown in Fig. 3.17. Initially there is an elastic strain (and, perhaps, a plastic strain). This is followed by a period of deformation in which deformation occurs at a steadily decreasing rate (stage I; primary or transient creep). This is followed by a secondary, steady-state creep regime (stage II), which in turn leads to a region where the creep rate increases and in which the strain may become localized until rupture occurs (stage III; tertiary or accelerating creep).

The steady-state creep stage is particularly sensitive to both stress and temperature such that the strain rate, $\dot{\varepsilon}$, increases with both stress and temperature through a relationship such as:

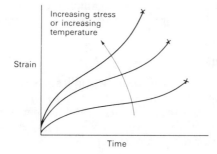

Strain

Increasing stress
or increasing
temperature

Time

Fig. 3.17 A family of creep curves showing the effect of increasing stress or increasing temperature

$$\dot{\varepsilon} = A\sigma^n \exp\left(\frac{-Q}{RT}\right)$$

(3.15)

where A and n are material constants and Q is an activation energy.

During the steady-state stage there is a balance between work-hardening effects and recovery effects. Subsequently there is extensive void formation, predominantly at grain boundaries, during stage III, and the final fracture is produced by the linkage of these voids.

At higher temperatures ($> 0.9T_m$) an additional mode of deformation known as grain-boundary sliding occurs.

Typically creep testing involves loading at constant stress and temperature for very long periods of time. In practice, materials subject to loading conditions that might give rise to creep might be reasonably expected to exhibit creep lives of 20 years or more. Clearly it is difficult to assess behaviour for design purposes over such an extended test period. Accordingly it is very common to make use of *stress–rupture* testing. In a stress–rupture test the time to cause failure at a given stress and temperature is measured under stress conditions where failure might be expected in 1000–10 000 hours. Stress-rupture lives at lower stresses and temperatures are then predicted from these data using such equations as that for the Larson–Miller parameter, L, given by:

$$L = T\,(20 + log_{10}t)10^{-3}$$

(3.16)

where t is the time to failure.

In service applications, it must be recognized that the component undergoing creep, e.g., a gas turbine blade, will be subjected to both variations in temperature and stress which makes accurate predictions of performance extremely difficult.

Metallic alloys are the most widely used creep resistant materials and these are required to exhibit both resistance to oxidation and microstructural stability. Although most alloys have been developed empirically, increasing understanding of behaviour during creep has led to substantial advances in alloy design. The most significant of these have been the nickel-base superalloys now produced to meet a wide range of specifications.

3.2.2 Physical properties

Although in most structural *engineering* applications mechanical properties are of great importance, it is necessary also to have knowledge of various physical properties that can affect in-service performance. For non-structural applications, physical properties, e.g., magnetic and/or electrical properties, can be of primary importance.

Density
    The density of a material specified in terms of its weight per unit volume is determined by the atomic weights of the constituent atoms and their packing arrangements. Predictions for metals and alloys based upon compositional and crystal structure data correlate closely with measured values which largely lie within the range 3000–10 000 kgm^{-3}. Non-metallic materials such as ceramics and glasses tend to have densities at the bottom end of this range while polymer densities are commonly in the range 100 kgm^{-3} (foamed polystyrene) to 1000 kgm^{-3} (polymethylmethacrylate).

Bulk densities are of importance because they give rise to self stresses in structures. These self-stresses can vary from dead-weight bending or uniaxial stresses to the centrifugal stresses occurring in rotating components. Unfortunately, however, density is a structure-insensitive property and comparative improvements in behaviour must be achieved by careful materials selection rather than by modifications in processing or heat-treatment.

When considering *specific* performance it is important to take into account the stressing system. Thus in the design of minimum weight structures the important parameter for tensile loading is the modulus/density ratio, E/ϱ^1, whereas for resistance to bending or buckling the corresponding parameters are E/ϱ^2 and E/ϱ^3, respectively.

Coefficient of thermal expansion
    When the temperature of a metal or alloy is raised, the atoms undergo increased thermal vibration which leads to an expansion of the lattice. The resulting strain increment is given by:

$$\Delta\varepsilon = \alpha_T \cdot \Delta T \tag{3.17}$$

where α_T is the coefficient of thermal expansion and ΔT is the range of temperature change. For most metals and alloys α_T lies in the range 10–20 $\times$ 10^{-6} K^{-1}. Changes in temperature can give rise to very substantial *thermal stresses* through the consequent dimensional changes. These must normally be taken into account in design considerations. When non-uniform temperatures occur or when a material is exposed to a temperature cycle, e.g., during a process such as welding, the resulting conditions may indicate that there is no net stress or strain although there may be significant point-to-point variations in internal stresses and strains. These so-called *residual* stresses are additive to the service stresses and can lead to premature failure.

Electrical properties

The electrical conductivity is by far the most important electrical property of metals and alloys. Other electrical properties, such as dielectric properties, piezoelectricity and thermoelectric performance, are also of importance in special applications.

The electrical conductivity is usually defined as the reciprocal of the electrical resistivity, ϱ, which is related to the electrical resistance, R, by:

$$R = \frac{\varrho L}{A}$$

$$(3.18)$$

where L and A are the length and cross-sectional area, respectively, of the conductor under consideration. Thus, the resistivity is given in units of ohm.m. For a given material the conductivity is a product of the number of charge carriers, the charge per carrier and the mobility of each carrier. The conductivities of metals and alloys are high ($\sim 10^7$ $(ohm.m)^{-1}$) which reflects their high density of mobile, charge-carrying electrons. However, alloying can markedly reduce conductivity, whereas processing effects, e.g., cold-working, are of lesser significance.

One special class of materials in electrical applications is the semi-conductors such as silicon and germanium. Semiconductor behaviour can either be described as intrinsic when conduction is promoted by thermal activation, or extrinsic when impurities or dopants are introduced to produce controlled numbers of charge carriers of specified types (n-type if conduction is through negatively-charged electrons or p-type if conduction involves positively-charged 'holes'). Combinations of n-type and p-type semiconductors are used as the building blocks of a wide range of electronic devices.

Magnetic properties

The important magnetic properties are the permeability, the saturation induction, the remanent induction and the coercive magnetic force. These are characterized by observing the relationship between the magnetic flux density, B, and the magnetic field strength H. This relationship in the form of the B-H curve is as important in describing magnetic properties as is the stress–strain curve in describing mechanical properties. By following a complete cycle of magnetization an hysteresis loop can be generated (Fig. 3.18) which can be used to identify the properties specified above including the permeability μ ($=B/H$), which varies with field strength.

For metallic materials *ferromagnetism*, in which the relative per-meability (the ratio of the permeability to the permeability of a vacuum) is high, is of greatest engineering importance. Ferromagnetism occurs in only a few metals, e.g. iron, nickel and cobalt. These form the basis of a range of magnetic materials that find use in transformers, electric motors, etc.

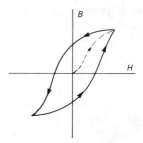

Fig. 3.18 A magnetic hysteresis loop obtained by determining the relationship between magnetic flux density, B and magnetic field strength H

Ferromagnetic materials are usually divided into *soft* magnetic materials, in which the area of the B-H loop (the magnetic hysteresis loss) and the saturation magnetization are small, and *hard* magnetic materials which have high remanent magnetization and coercive field. Most usually the best hard magnets are complex metallic alloys, although more recently non-metallic materials, generally called ferrites, have assumed a dominant role.

One special consideration with soft magnetic materials such as the iron–silicon alloys widely used in transformer cores is the anisotropy of magnetization of the basic crystal structure (Fig. 3.19). By controlled processing the softest magnetic direction, <100>, can be located so as to produce sheet with very low-loss properties.

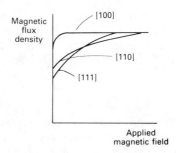

Fig. 3.19 The anisotropy of magnetisation in body-centred cubic iron obtained by comparing the response of crystals with [100], [110] and [111] axes, respectively

3.3 ANISOTROPY OF BULK PROPERTIES

It is normal to consider engineering materials to be isotropic. Thus, representative values of properties, such as the elastic modulus or yield stress, are usually presented in lists of reference data. Isotropy is, however, rare in practice. Most metallic solids after forming are substantially anisotropic, in that they exhibit differences of properties in different directions. The mechanical and physical properties of crystalline materials are a function of both the properties and of the individual phases and the pattern of their arrangement.

In a metallurgical sense, anisotropy arises from two principal sources:

(i) crystallographic anisotropy, in which the crystal grains in a polycrystalline aggregate are arranged in a non-random fashion to give a preferred orientation or texture

(ii) microstructural anisotropy, in which there is substantial variation in grain size and shape and the microstructural constituents are arranged in a non-random way. This latter form of anisotropy is most marked in materials such as steels that exhibit ferrite–pearlite banding (Fig. 3.20). Although engineers usually require isotropy, it is becoming increasingly clear that controlled anisotropy can frequently be very advantageous. For example, controlled aniso-tropy can be used:

(a) to give directional strength
(b) to give improved formability
(c) to produce specific in-service properties, such as with textured silicon–iron of low magnetic hysteresis loss for use in trans-former cores

It is well known that for most single crystals the mechanical and physical properties are orientation-dependent. If the aggregate of crystals has a true average random orientation and there are a sufficiently large

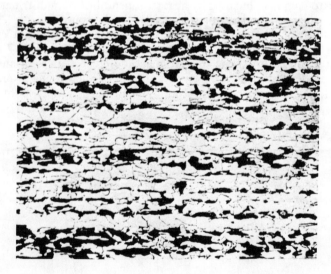

Fig. 3.20 A banded structure in a hypoeutectoid steel: magnification, 100×

number of crystals then the average properties of the aggregate will be isotropic. If texture exists this will not be the case.

The study of anisotropy can be broken down into several steps. Crystallographic anisotropy is studied either directly by measuring the preferred orientation using X-ray techniques or electron microscopy or indirectly by making measurements of mechanical and/or physical properties that reflect the texture. Microstructural anisotropy of grain shape and size and phase distribution can be assessed using the many methods of quantitative metallography currently available.

Texture is normally described by linking together a crystallographic specification with a feature of the specimen geometry. For instance, texture in wires and rods can be related to the wire or rod axis by attributing to the material a fibre texture $<uvw>$. This means that crystallographic directions of the type $<uvw>$ lie parallel to the axis. Sheet textures are described as being of the type $\{hkl\}<uvw>$. Then $\{hkl\}$ planes can be considered to lie in the rolling plane with $<uvw>$ directions parallel to the rolling direction.

Anisotropy of grain size and grain shape can be a consequence of segregation affecting recrystallization or inhomogeneous deformation and of differences in thermomechanical treatment at various sections through the thickness. It is particularly significant if the material shows a marked grain size dependence of the flow stress. Then hard and soft regions will occur as the grain size varies.

The common forms of phase distribution anisotropy are microstructural banding (Fig. 3.20) and the formation of inclusion stringers, such as those observed, in an extreme case, in wrought iron. In both cases quantitative metallography can be used to specify the degree of anisotropy by yielding information on the shape and distribution of the phases. It is usually possible to eliminate banding by heat treatment. Inclusion distributions are more troublesome and are best controlled at the source.

This form of anisotropy commonly has the most significant effect on transverse ductility and transverse fracture toughness. Table 3.7 presents data for a forged low-alloy steel in which different degrees of anisotropy have been developed by progressively increasing the forging reduction.

As referred to in the previous section, the magnetic properties of iron are significantly anisotropic, and this can be used to advantage.

Table 3.7 Directionality of properties in low-alloy steel forgings

Reduction ratio	Direction tested	Tensile strength (MNm^{-2})	Elongation (%)	Impact strength (J)
1.7	Longitudinal	800	20.0	66
	Transverse	800	18.0	54
3.2	Longitudinal	800	20.0	80
	Transverse	793	16.0	39
6.1	Longitudinal	793	22.0	98
	Transverse	793	12.0	34

Mechanical properties such as the yield stress and Young's modulus can also vary significantly in metals and alloys with strong preferred orientation. For example, the angular variation of the proof stress in the plane of a textured sheet can vary by up to about 10%. More marked variations in Young's modulus can occur in textured sheets of metallic materials with anisotropic single crystal properties such as iron or copper. Then variations of up to 30% are observed.

3.4 SURFACE PROPERTIES

Although it is clear that the bulk properties of solids are of great significance in determining their performance in service, there are many instances when surface properties have an important role. This is particularly the case where wear resistance and corrosion resistance are concerned.

3.4.1 Wear resistance

When the surface of a solid component moves in contact with another solid body various interactions can occur. These give rise to the two principal forms of wear, *abrasive* wear and *adhesive* wear (see Figs. 4.18 and 4.19). Both of these involve elastic and plastic deformation, whereas in the latter case there is a stage at which bonding occurs between the two surfaces in contact.

In considering wear behaviour, it must first be recognized that when two surfaces come together with only limited normal pressure they only make contact at a few elevated points. As the normal pressure increases, contact is increased. If one surface is much harder than the other, then usually it will only deform elastically while the softer surface deforms both elastically and plastically. If the surfaces are then moved relative to each other, the more easily deformed surface is deformed substantially and grooves are 'ploughed' in the surface by the asperities on the harder surface. This is abrasive wear. As the softer material work hardens, pieces of the surface may even become detached. Abrasive wear is made more severe if hard particles, such as sand or oxide, are entrapped between the rubbing surfaces. Then, depending on the relative hardnesses, both surfaces may be abraded. In order to avoid abrasive wear it is conventional to surface-harden the contact surfaces as well as both using an effective lubricant between them and ensuring that the system is free from abrasive particles.

Even in the presence of lubricants or thin surface oxide layers, it is possible for a breakdown in the surface coating to occur so that the two surfaces in contact adhere together. Then, if lateral displacement occurs, the softer material will be locally deformed and extended until a fragment breaks away from the surface. Thus, the harder surface becomes progressively covered with small pieces of the softer. This is adhesive wear. The bond is most usually diffusive and the adhesion can be quite severe.

Clearly as the wear process continues, the softer material becomes pitted as the fragments are removed.

If the contact surfaces are in oscillatory contact rather than rubbing continuously then surface fatigue can occur, leading to cracking and localized deformation. This can also occur during rolling contact, as in ball or roller bearings, because of the oscillatory nature of the surface contact stresses. If the nature of the contact conditions is such that periodic localized heating occurs, then the induced thermal stresses can hasten the accumulation of damage. Enhanced damage and wear can also result if chemical reactions occur in the interface region. This combination of wear by abrasion, adhesion and fatigue is known as *fretting* (see Fig. 4.20).

As with abrasive wear, adhesive wear and fretting are minimized by a combination of surface hardening, lubrication and environmental cleanliness.

3.4.2 Corrosion resistance

In the presence of aggressive gaseous environments, e.g., hot exhaust gases, or aggressive aqueous environments, e.g., salt water, considerable and irreversible surface damage can occur. The responses are known as oxidation and corrosion, respectively. In order to preserve component integrity it is, therefore, essential to ensure that the surface exhibits resistance to the environment. Thus, corrosion resistance is required and although it is usual to consider dry and wet corrosion separately the basic principles involved in each case are essentially the same.

Obviously one simple approach is to remove the component from contact with the aggressive environment. This is why heat treatment is carried out in vacuum or in an inert or non-oxidizing atmosphere. More commonly, however, the component surface is provided with some form of protection either through a surface treatment or by surface coating. With aqueous corrosion special techniques, such as those involving cathodic protection or the use of corrosion inhibitors, can be used to minimize or eliminate the interaction between component and environment.

As an oxide is formed on a metallic surface, a barrier is created through which ionized atoms must pass in order for the oxide film to continue to grow. If the rate-controlling step is diffusion through the oxide, then the rate of thickening is parabolic. In other circumstances, linear or logarithmic growth behaviour are observed. With alloys preferential oxidation of component species can occur. Thus the presence of chromium, aluminium and/or silicon can impart oxidation resistance, since in each case a coherent, protective surface oxide film is produced. The use of 'selective oxidation' is very effective and is the basis of the oxidation resistance of the stainless steels and the creep-resistant nickel-base alloys.

In aqueous environments the most effective methods of gaining corrosion resistance involve modification of the properties of the parent metal or alloy, or the use of protective coatings. The principles underlying modification of properties are similar to those involved in producing oxidation resistance. A primary aim is to produce some form of surface film that is stable in the corrosive environment (be it acidic, neutral or alkaline) and which affords protection. Thus, with the stainless steels, this is achieved by producing a passive chromic oxide surface film. As knowledge of corrosion mechanisms has been gained, it has become possible to design alloys to obtain acceptable behaviour under different conditions. For example, it is necessary to add a small amount of molybdenum to stainless steels to produce the required performance in acidic environments.

Protective surface coatings can be either metallic or non-metallic. Here the aim is to produce changed surface properties. One of the simplest procedures is to apply a corrosion-resistant surface coating to the material prone to corrosion. This can be done in various ways, such as by cladding, dipping, spraying, electrodeposition or cementation. It is common practice to deposit surface coatings of aluminium, zinc, tin or chromium using these processes. Alternatively inorganic coatings, such as phosphates, anodized oxides and glassy enamels, or organic coatings, such as polymeric resins, oils and bitumen, can be applied to the surface. In some cases, as with paints, a combination of metallic and non-metallic components is used. Surface coatings are usually effective without having an effect on the intrinsic properties of the coated artefact. This is undoubtedly one of the reasons why they are so widely used. However, it is important to recognize that they provide effective protection only while they are undamaged. In many cases some short-term resistance to minor damage can be achieved but in the longer term repair and/or replacement of the surface coating is necessary.

In general it needs to be recognized that it is not necessarily the amount of corrosion that is of greatest importance but its distribution. Thus localized attack can have very damaging effects, even though the total amount of damaged material is small.

3.5 THE INTERACTION OF SERVICE STRESSES AND STRUCTURE AND THE INFLUENCE ON PROPERTIES

When a metallic component is in use, it is principally subjected to the applied working stresses. In addition there are possible secondary stresses as a consequence of self loading or residual stress. Very sophisticated procedures are available for analysing the working stresses in complex components under various forms of loading. These analyses take into account both elastic and plastic behaviour, and it is usually assumed that the material from which the component is formed is homogenous and isotropic. It is quite feasible, however, for both elastic and plastic anisotropy to be allowed for. As a result it is not common for service failures to arise because of inadequacies of the analyses of service stresses. However, the existence of external and/or internal discontinuities of shape or structure can perturb the stress field. These perturbations are usually quite localized. As a result localized failure can occur. Further, it is possible for these localized stress perturbations to give rise to conditions wherein localized structural changes occur so that the material properties are altered. For example, localized plastic shearing of precipitation-hardened aluminium alloys can lead to reversion (dissolution of the precipitates) and local softening. This further concentrates the deformation and catastrophic failure follows.

A common source of localized stress concentration is a change in section. Additionally microstructural non-uniformities such as inclusions can have a deleterious effect. Sometimes microstructural non-uniformities can be produced during manufacture and then retained. This is also the case with residual stresses that arise during manufacture. During manufacture elastic and plastic deformation occur and although the manufacturing loads are removed, the resulting springback can induce stress. Thus, after bending or torsion a complex stress pattern exists in the formed component. In summary, the total stress σ_{total}, is given by:

$$\sigma_{total} = \sigma_{applied} + \sigma_{sc} + \sigma_{residual} \qquad (3.19)$$

where $\sigma_{applied}$, σ_{sc} and $\sigma_{residual}$ are the contributions from the applied stress, stress concentration and residual stresses, respectively.

Stress concentrations associated with changes in section or notches or cracks can be analysed by using elasticity theory. For a surface step or a notch, the magnification of stress associated with the feature is dependent on the root geometry. The stress concentrating effect increases with decrease in root radius as shown in Fig. 3.21. This is also the case for internal cracks (Fig. 3.22). As a consequence atomically sharp cracks, such as those produced by fatigue or cleavage, e.g., quench cracks occurring during heat treatment, are particularly dangerous.

Many microstructural defects such as isolated, elongated inclusions (Fig. 3.23a) or clouds of inclusions (Fig. 3.23b) can act as internal cracks having stress concentrations associated with them. The material performance is then downgraded and premature failure can result. Fatigue resistance is one of the most significantly affected properties (Fig. 3.24).

It is, therefore, very important to recognize that the various factors associated with producing and manufacturing metallic articles can impose severe limitations on the achievement of the required in-service properties.

3.6 LIMITATIONS IMPOSED ON THE ACHIEVEMENT OF PROPERTIES

3.6.1 Effects of fabrication processes

The production of an engineering component usually involves casting of the base material followed by some or all of a series of working, forming and machining processes. In addition special treatments, such as case hardening or surface coating, can have substantial effects. Thus it is appropriate to examine each of these production stages in order to recognize the nature and origins of the associated microstructural and macrostructural defects with a view to identifying methods of eliminating them or of minimizing their effects.

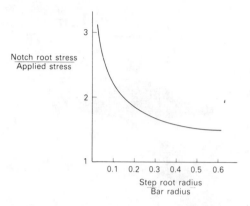

Fig. 3.21 The stress concentration associated with a surface step in a bar under tension

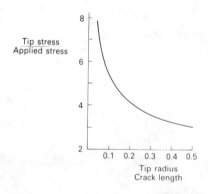

Fig. 3.22 The stress concentration associated with an elliptical internal crack in a bar under tension

Casting

As a process, casting can be used to produce ingots or continuously-cast strands for subsequent working to finished or semi-finished stock or for the direct production of castings for use after a minimum amount of dressing and machining. The structure of a cast material is of great importance since many materials properties, especially mechanical properties, depend on grain shape and grain size. In addition, segregation resulting from the various modes of redistribution of dissolved alloying elements can have marked effects. In most as-cast polycrystalline solids three distinct zones with different grain structures can be identified (Fig. 3.25):

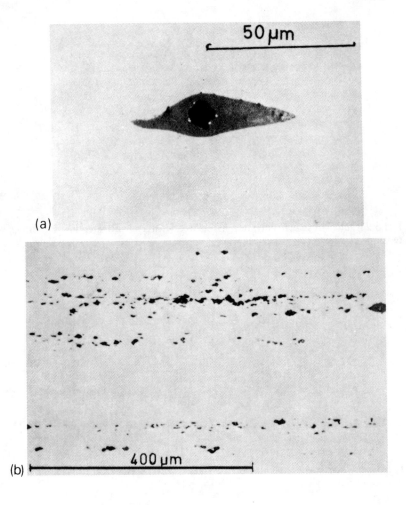

Fig. 3.23 (a) An elongated MnS inclusion (with a calcium aluminate core) in a hot-rolled steel. (b) Clusters of Al_2O_3 inclusions in a hot-rolled steel

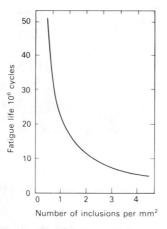

Fig. 3.24 The variation of fatigue life with the number of inclusions per mm^2 in a bearing steel

(i)   the chill zone — a boundary layer, adjacent to the mould wall, of small equiaxed crystals with random orientations
(ii) the columnar zone — a band of elongated crystals aligned parallel to the directions of heat flow
(iii) the equiaxed zone — a central region of uniform crystals. The properties of this central region are comparatively isotropic provided the grain size is small. The grain size in the equiaxed zone is, however, normally larger than that of the chill zone.

Probably the most important factor in determining subsequent properties is the relative proportions of the columnar and equiaxed zones. The chill zone is normally only a small number of grains thick and has a very limited influence.

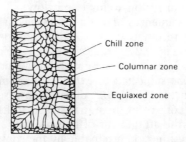

Fig. 3.25 Grain structure of an as-cast ingot (schematic) showing chill, columnar and equiaxed zones

If the object of the control of grain structure is to obtain isotropy, then a fine-grained equiaxed structure is required. This is brought about by

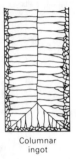 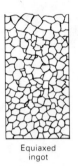

Columnar
ingot

Equiaxed
ingot

Fig. 3.26 Comparative grain structures of a fully columnar ingot and a fully equiaxed ingot

encouraging those conditions which lead to the breakdown of columnar growth and the formation of the equiaxed zone. On the other hand, if anisotropic properties are required, the columnar zone will need to predominate. The relative proportions of the different zones can be controlled by altering the casting variables, for example, the alloy composition, the pouring temperature, the rate of cooling and so on, and an ingot can vary from being fully columnar to being fully equiaxed (Fig. 3.26).

The grain boundaries form by the impingement of the growing grains and the effects of the boundaries derive both from crystallographic sources and particularly because of the accumulation of soluble and insoluble impurities that can occur. Thus, the effects of these grain boundaries depend both on the change in orientation at the boundary and on the special properties of the boundaries themselves. So far as fracture is concerned, they can form easy paths for crack propagation, especially in the presence of solute segregation. If the equiaxed zone is absent, accumulation of soluble and insoluble impurities in the regions where the columnar structures impinge can have disastrous results; for example, this is commonly the case in fusion welds (see below).

In castings, the consequences of grain impingement are particularly bad at sharp corners, in rectangular sections and at perpendicular surface junctions, as shown in Fig. 3.27. These can usually be corrected by suitable design procedures. In considering these it should be noted that if the ingot structure is going to be subjected to *extensive* working in the solid state, the final grain structure will be dictated by solid state changes. Nevertheless, careful control in the initial stage can have beneficial results particularly as far as the attainment of homogeneity is concerned. This is an important ancillary function attainable by the control of grain size and shape.

Segregation results from the various ways in which the solute elements can become redistributed within the solidified structure. The different types of segregation are usually classified as microsegregation or macrosegregation. Microsegregation is a short-range phenomenon and

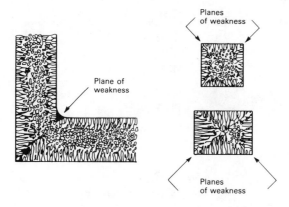

Fig. 3.27 Planes of weakness association with grain impingement in the corners of a rectangular section and at a perpendicular surface junction

extends over distances of the order of the grain size or less. In the case of cellular or dendritic segregation, the compositional differences may be confined to distances of a few microns. When the compositional differences show long-range variation, for instance between the outside and inside of an ingot, this is considered as macrosegregation. It is normally considered to occur over distances of greater than a few grain diameters. To a large extent the compositional variations that occur adjacent to the solid–liquid interface during solidification determine the nature and extent of segregation. For macrosegregation the longer range flow of unsolidified liquid is of great importance.

Superimposed on the cast structures of ingots discussed in the previous chapter are macroscopic segregation patterns that for ingots show both horizontal and vertical variations. The most characteristic of these is the segregation pattern of sulphur in cast killed-steel ingots. This is shown in Fig. 3.28. The regions of positive (concentration greater than average) and negative (concentration less than average) segregation are as indicated.

The effects of segregation are threefold:

  (i) the variations in composition can produce variations in composi- tionally-dependent properties
 (ii) the variations in composition can lead to differences in heat treatment response from point to point, which in turn give rise to property variations
(iii) undesirable intermetallic phases and inclusions can form in regions where the composition varies significantly from the prescribed composition.

The production of sound (defect-free) castings is of great economic importance. It is an essential step to be able to recognize the different kinds of defects that can arise in casting processes, and when they occur to

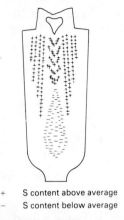

+   S content above average
−   S content below average

Fig. 3.28 Typical segregation pattern for sulphur in a killed-steel ingot

define the steps necessary to eliminate them from subsequent castings. Design and practice are of importance in influencing the incidence of casting defects. Often errors in design can be overcome by the adoption of procedures involving careful control of the cast metal and of the casting procedure. In many cases, in order to facilitate production it is of greater practical and economic merit to modify the design of an object to be cast. Casting defects are of six principal types:

(i)   *blowholes* — these are round or elongated cavities, usually with smooth walls, found on or under the surface of castings
(ii) *cold-shuts* — these are produced when two streams of metal flowing from different regions in the casting meet without union
(iii) *contraction cracks* are irregularly shaped cracks formed when the metal pulls itself apart while cooling in the mould or after removal from the mould
(iv) the development of *'flash'* (or *'fin'*) involves the formation of a run of metal around the parting line of the mould caused by the flow of liquid metal into the space between the halves of the mould
(v)   *oxide and dross inclusions* result from the entrapment of surface oxide or other foreign matter during pouring
(vi) *shrinkage cavities* — these formed because of the change in volume that occurs on solidification which, in isolated regions can give rise to a cavity of irregular shape.

The causes of casting defects and both foundry and design remedies are well documented. Their elimination is an important step in the achievement of the properties specified in engineering design.

Working and forming
Primary working processes such as forging, rolling, extrusion, wire and

rod drawing and tube making can also give rise to defects and microstructural and macrostructural features which seriously affect properties. Again their causes and remedies are well understood. However, this does not mean that their occurrence is rare. Control of production processes can be difficult and the reliable maintenance of quality is a goal always aimed for. Nevertheless, it is very much a consequence of improvements in inspection and assessment that has led to significantly reduced incidences of defective material in service. The various forms of defects that occur are summarized in Table 3.8.

Table 3.8 Common defects in worked products

Working process	Major defects	Minor defects
Forging	Surface and/or flash cracking Cold shuts Internal cracking	Incomplete penetration Buckling Twinning
Rolling	Surface marks Variations in dimensions Barrelling Edge/surface cracking	Internal fissures 'Alligatoring'
Extrusion	The extrusion defect 'Fir-tree' cracking Variations in structure	
Wire, rod, tube drawing	Internal cracks Surface marks	Discoloration Slivers and seams

One of the more marked manifestations is mechanical *fibring* which is a form of microstructral anisotropy (see also Section 3.3) in which microstructural constituents are aligned and elongated in directions determined by the flow of metal. Often the resulting 'flow lines' can be used to advantage because the mechanical properties tend to be superior along the fibre direction, whereas crack propagation transverse to the flow lines is much more difficult than propagation parallel to them. An example of a desirable flow line distribution in a forging is shown in Fig. 3.29.

Another manifestation of working is the change in inclusion shape and distribution that can occur. Inclusions which are ductile at working temperatures, e.g, MnS (cf.Fig. 3.22a), can elongate to form ideal regions for crack nucleation. Brittle inclusions can break up and leave what are effectively cracks in the matrix.

Working and forming processes are usually carried out hot, at temperatures well above $0.5T_m$, or cold. In the former case recovery and recrystallization processes occur during deformation. It is usual after cold-working to subject the material to an annealing treatment to promote recovery and recrystallization. In either case the response is dependent both upon the composition and the imposed strain. For instance, the grain

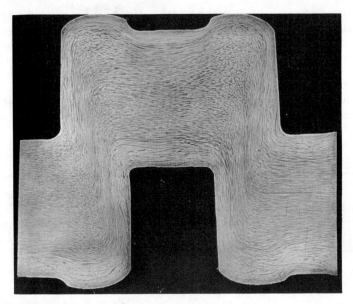

Fig. 3.29 Flow lines in a steel forging

size after recrystallization is dependent on the strain during working, such that once the critical strain ($\sim$3%) is exceeded, the recrystallized grain size decreases with increasing strain. It follows that different regions that have been subject to different strains will exhibit different grain sizes after recrystallization. This can have an influence on properties. For example, in the cross-section of a hot extruded bar in which there is a strain gradient across the bar, this is reflected in grain size variations across the diameter.

Forming processes can also give rise to defects that affect in-service performance. This is particularly so for sheet-formed parts in which a spectrum of defects occurs such as stress cracking, wrinkling, stretcher strains, earing and 'orange peel'. Some of these defects have more effect on the appearance of the finished part than the performance but notwithstanding this they should be avoided.

Machining

Machining is often carried out on cast or wrought components in order to achieve the required dimensional accuracy and surface finish. There are a great many possible machining operations that remove material at a range of cutting speeds and depths. Machining processes involve local plastic deformation and fracture. This can leave behind surface and subsurface damage. Further, since there is usually a substantial amount of local heat generated there can be local microstructural modifications. These can vary from local tempering of hardened steels to local overageing of precipitation-hardened aluminium alloys. The machining process can

Fig. 3.30 A fatigue fracture initiated at a machining mark in the fillet of a steel crankshaft

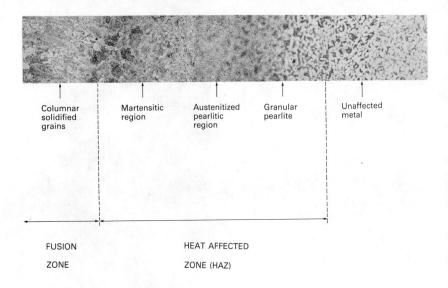

Columnar   Martensitic   Austenitized   Granular   Unaffected
solidified   region   pearlitic   pearlite   metal
grains   region

FUSION   HEAT AFFECTED

ZONE   ZONE (HAZ)

Fig. 3.31 The microstructure of the fusion zone and heat affected zone of an arc-welded steel. The principal microstructural constituents are indicated

also yield a roughened surface that essentially constitutes a series of micro-notches. All of these surface effects can make it difficult to achieve required properties. In particular machining marks can provide initiation sites for fatigue cracks (Fig. 3.30). As a result, care should be given to remedial treatment after machining to ameliorate the undesirable consequences of the machining process.

Special treatments

Frequently engineering components are given special treatments to develop particular properties for a specific purpose. These may range from general heat treatments, such as quenching and tempering, to localized treatments, such as surface hardening. In both situations, modifications to the microstructure and macrostructure may occur that limit service performance. This is less often the case with general heat treatments, although even then some difficulties can be encountered. For instance, if the component being treated has a complex geometry involving changes in section size and thickness, then differential contraction during quenching can produce quench cracking. This commonly occurs where surface steps or notches exist, and is encouraged by the associated stress concentration. Although stress relaxation occurs during subsequent tempering the cracks remain and these can provide sites for brittle fracture or fatigue fracture initiation. The effect on fatigue performance can be very marked, since the need to nucleate the fatigue crack is removed.

More general difficulties can also occur. One notable example is temper embrittlement. This occurs in many alloy steels if they are tempered in the range 500–600°C or allowed to cool slowly through this range. It has been established empirically that small additions of molybdenum can eliminate this problem.

As described earlier, surface treatments are used to provide wear resistance and corrosion resistance. During surface hardening by carburizing or nitriding followed by induction hardening, the aim is to ensure that the surface is subjected predominantly to compressive stresses. If, however, the hardening is non-uniform, then local stress variations can produce cracking. Conditions that promote cracking can also occur during coating treatments. It is observed that electrodeposition processes using aqueous electrolytes involve some hydrogen evolution. With ferrous materials, some of this hydrogen can diffuse into the surface and give rise to hydrogen embrittlement.

As a general rule it is advisable to make very careful quality assessments in all cases where materials have been given special processing treatments.

3.6.2 Effects of joining processes

During the assembly of machines and structures it is usually necessary to join together the different parts that make up the assembly. Joining can involve either a permanent join by welding, brazing or soldering, or demountable fastenings such as bolts and rivets. Both permanent and demountable joints can have harmful effects. The most significant effects occur during welding, and these warrant most detailed attention.

Welding

Welding procedures are divided into two broad categories. *Fusion* welds involve localized heating in the vicinity of the joint to above the parent metal melting temperature with or without the separate addition of molten filler metal. In *pressure* welding, on the other hand, no general melting occurs and the bond is formed in the solid-state (at most there may be some very localized melting of surface asperities). It is the thermal cycle associated with these welding processes that is responsible for changes in the weld metal and in the heat affected zone adjacent to it. The microstructure of the fusion zone and heat-affected zone is complex and contains a range of constituents (Fig. 3.31). Because of the higher temperatures involved, the defects that occur in fusion welding are usually the more deleterious.

Fusion-welding processes are classified according to their source of heat, with subdivisions related to the method of shielding employed. Pressure welding can take place hot or cold, although hot pressure welding is more common. Again the source of heat is an important characteristic. The principal processes are listed in Table 3.9.

After melting and solidification, the fusion zone exhibits a typically cast structure with predominantly columnar grains. All of the common casting defects can occur, and it is necessary to control conditions carefully to

Fig. 3.32 Cross-section of a butt weld showing the columnar grain structure in the fusion zone

Table 3.9 Fusion- and pressure-welding processes

Fusion	
Manual metal arc	Metal inert gas (MIG)
Tungsten inert gas (TIG)	Submerged arc
Thermit	Electroslag
Gas	Electron beam
Plasma	Laser
Pressure:	
Resistance	Spot
Hot pressure	Cold pressure
Flash butt	Friction
Explosive	Diffusion

avoid gas porosity, shrinkage cavities, entrapped inclusions, etc. Further, because of the columnar growth centre line, segregation of impurities frequently occurs. A typical weld cross-section is shown in Fig. 3.32. The geometry of the weld bead can give rise to stress concentrations. The thermal cycle leads to the occurrence of residual stresses. These predominate in or around the heat-affected zone. Rapid heating and cooling often produces martensite in the heat-affected zone close to the fusion zone boundary. The presence of martensite can lead to cold cracking of the parent plate (Fig. 3.33). An important variable in determining the cracking susceptibility of steels is the carbon equivalent value (CFV) given by:

$$CEV = \frac{Mn}{6} + \frac{Cr + Mo + V}{5} + \frac{Ni + Cu}{15} \qquad (3.20)$$

If this exceeds about 0.45, then preheating of the plates to be welded and post-welding heat treatment are likely to be necessary. Cold cracking is aggravated by dissolved hydrogen produced by the break-down of atmospheric moisture. Thus, it is recommended that steps are taken to exclude moisture from the vicinity of the weld pool during the welding process.

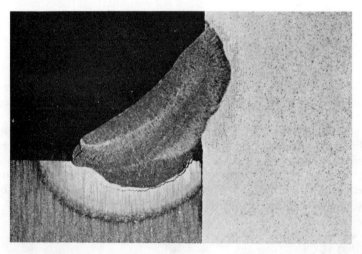

Fig. 3.33 Cold crack in the heat affected zone adjacent to a fillet weld in steel

For critical components and structures extensive non-destructive testing is carried out after welding to ensure freedom for defects. If defects are found they are cut out and rewelding undertaken. In addition careful *in situ* heat treatments are carried out in order to optimize the properties of the weldment.

With pressure welding, although the thermal cycle is less severe, it is still

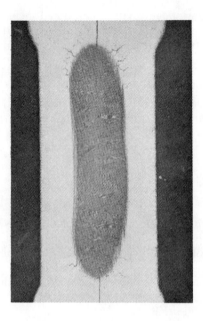

Fig. 3.34 Localized cracking in the zone adjacent to a spot weld
between nickel-alloy sheets

possible for residual stresses and cracks to occur (Fig. 3.34). Pressure
welds are, however, less likely to be used in critical load-bearing
applications.

   In all cases the intention is to produce a welded structure with properties
as similar to those of the parent material as possible. This ideal is rarely
achieved. In most circumstances the welded area exhibits properties close
to but marginally below those of the parent material. Thus it is important
to recognize that the use of welding does offer a limitation to the
achievement of specified properties.

Fasteners
   The use of fasteners, e.g., bolts and rivets, does not in its own right
usually impose limitations on material behaviour. However, there is one
important circumstance when attention is necessary. This is when the joint
is exposed to a corrosive aqueous environment or when condensation is
likely to occur. Then regions away from the surface can become strongly
anodic so that localized pitting and/or crevice attack occurs. When there is
dynamic loading fretting corrosion is likely in bolt holes or rivet holes.
This latter form of corrosion is particularly insidious, since it is almost
undetectable until serious damage has been produced.

3.6.3 Effects of operating conditions

Load variation

In most cases load variations do not have a significant effect. There are two main exceptions. The first and more important is under fatigue conditions. The second is during creep.

Most fatigue data are obtained by assessing performance during cyclic loading. For instance, stress-life testing identifies the number of cycles to failure for different loading ranges. In practice, components are loaded so that the load range varies throughout the lifetime. Then it is usual for the occurrence of cumulative fatigue damage. Thus if a loading history involves n_1 cycles at a stress amplitude $\Delta\sigma_1$ where the fatigue life is N_{f_1} followed by n_2 *cycles at a stress amplitude* $\Delta\sigma_2$ where the life is N_{f_2}, etc., the requirements is:

$$\frac{n_1}{N_{f_1}} + \frac{n_2}{N_{f_2}} + \ldots \quad \leqslant 1 \qquad (3.21)$$

This is Miner's rule and for most design purposes it is sufficient to ensure that the sum of the fractions of the lifetimes is unity. However, there is evidence to support the view that damage accumulates more rapidly at high stresses than Miner's rule predicts. Thus, in a situation where occasional, severe stress spikes are experienced, a substantial reduction in fatigue life can result. Then it is advisable for component testing to be carried out using simulated service conditions.

Similar considerations apply under creep conditions. This is especially so since the accumulation of creep cavitation is very stress dependent, particularly towards the end of the steady-state region. Thus, void nucleation and growth can be significantly increased during short periods of time under enhanced load. This effectively reduces the available life at lower loads and gives rise to what is described as premature failure.

Temperature variations

At temperatures below $0.5T_m$ when creep is not expected, the effect of temperature variations is to produce cycles of thermal expansion and contraction. This can lead to failure by thermal fatigue. This frequently appears as an array of small cracks on the surface known as 'thermal crazing'. This is particularly bad if sudden but severe thermal pulses are imposed, as in gun barrels or tool steel casting dies.

Microstructural stability is also important if temperature variations are not to lead to a deterioration of materials properties. Since most of the processes that bring about microstructural changes, e.g., over-ageing of precipitation-hardening alloys, tempering of quenched steels, are thermally activated, exposure at elevated temperatures, even for quite short times, can be damaging. If 'hot spots' occur then the breakdown may be very localized. This can at best produce undesirable dimensional changes, whereas at worst local failure can be initiated.

Creep processes are also thermally activated and, therefore, even quite short-term exposure at temperatures above the design temperature should be avoided.

So far, the effects of elevated temperatures have been considered. For steels in particular, exposure to low temperatures must not be overlooked. As described in Section 3.2.1, steels are prone to a ductile-to-brittle transition at temperatures below normal ambient temperatures. Even if catastrophic failure does not result (and frequently it has done!) cracking may occur that will reduce markedly the effective life of the structure or component.

Interactions with the environment

In Section 3.4.2, the need for oxidation and corrosion resistance was emphasized. There are, however, two hazardous conditions that occur when there is a combination of a corrosive environment and either a steady or a fluctuating stress. The failures that then occur are described as stress corrosion cracking or corrosion fatigue, respectively. Different alloys vary substantially in their responses to combinations of stress and corrosion. Those that fail usually do so without any gross corrosion taking place.

Stress corrosion cracking takes place at stress intensities very much less than those normally required for crack propagation. Thus a structure that is safely able to support the design load when first constructed becomes progressively unsafe with time. The time to failure is stress-dependent, as shown in Fig. 3.35. A decrease in stress is accompanied by an increase in lifetime. Further, there appears to be a threshold stress below which failure does not occur. This threshold stress does not, however, have any simple relationship to basic mechanical property parameters, other than it is usually the case that an increase in yield stress or tensile strength in a stress-corrosion-susceptible alloy produces a decrease in the safe threshold stress. Residual stresses are known to play an important part in stress-corrosion cracking and this requires that formed parts are annealed or stress relieved as immediately after forming as is practical.

Stress-corrosion cracking occurs when the loading is static. Corrosion fatigue occurs when the loading is dynamic. An important feature of corrosion fatigue is the enhancement of both crack initiation and crack propagation. During corrosion fatigue the crack propagates through

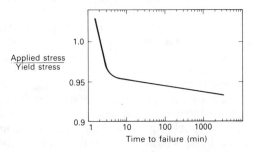

Fig. 3.35 The time to failure as a function of stress for the stress corrosion cracking of a magnesium alloy

conjoint action of mechanical and electrochemical mechanisms. The effects are most marked at stresses where the stress-corrosive contribution can play the greatest part.

SUGGESTIONS FOR FURTHER READING

J.M. Alexander and R.C. Brewer (1963) *Manufacturing Properties of Materials*, Van Nostrand Reinhold, Wokingham.

M.F. Ashby and D.R.H. Jones (1980) *Engineering Materials*, Pergamon, Oxford.

V.J. Colangelo and F.A. Heiser (1974) *Analysis of Metallurgical Failures*, John Wiley, New York.

J.P. Chilton (1972) *Principles of Metallic Corrosion*, Royal Society of Chemistry, London.

R.A. Flinn and P.K. Trojan (1981) *Engineering Materials and Their Applications*, Houghton Mifflin, Boston.

R.W.K. Honeycombe (1984) *The Plastic Deformation of Metals*, 2nd edn, Arnold, London.

J.F. Lancaster (1980) *The Metallurgy of Welding*, George Allen and Unwin, London.

CHAPTER 4

Performance in service

The failure of a component or a piece of equipment in service normally has far-reaching consequences. Not only is there the associated risk to human life but also there are severe economic effects. Not only can there be expensive replacement and repair costs but, in addition to these direct costs, there is also a consequential loss in production and possibly third-party liability damages to be met. As has been stressed in earlier chapters when designing equipment and selecting materials for service, the engineer and the metallurgist are faced with complex problems because of the interactions between processing, loading and environmental conditions. Most of the critical components used in aircraft, the chemical and petrochemical industry, and nuclear installations are designed and inspected according to rigorous specifications and codes of practice. Nevertheless failures do sometimes occur both during simulated life testing and in service. Much can be learned about performance in service by the analysis of such failures. From a knowledge of the possible contributory causes and mechanisms, any individual failure can be assessed and a diagnosis arrived at. In coming to an appropriate conclusion it is frequently valuable to be able to draw on case histories of the performance of components that had been subjected to similar conditions to the component under examination. The value of a successful diagnosis is substantial, since it provides a foundation on which the avoidance of subsequent failures can be based.

Some failures are sudden and catastrophic, others are not. A machine may fail prematurely during the earliest part of its working life, or it may be subjected to an unexpected hazard at sometime during its life, or it may slowly deteriorate and wear out. Of course, the more spectacular failures normally attract the most attention. Nevertheless, all failures deserve careful investigation, since the knowledge gained is most rewarding. Failure analysis can be systematic, but it must be recognized that the results obtained are of real value only if they are effectively disseminated. Sometimes manufacturers may not wish to share the information publicly because of adverse publicity. On other occasions, although there is no reluctance to share information, there may be serious inadequacies in the channels of communication such that those who may benefit from the information do not receive it. Engineers and others responsible for the selection of materials learn by their mistakes. Thus, on the basis of received knowledge of failure-diagnosis, appropriate changes can be initiated to increase product reliability.

The relationships that develop between the engineer and the materials specialist in design and materials selection are cyclic as can be illustrated in Fig. 4.1.

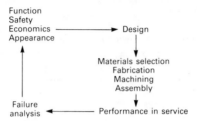

Fig. 4.1 A schematic relationship between design, service performance and failure analysis

In this chapter the causes of failure and a systematic approach to failure diagnosis will be considered. In a subsequent chapter the essential practice of quality assurance, the testing for defects during manufacture and prior to service, and the monitoring of components in service will be examined. Together these chapters provide a basis for maximizing the probability of effective performance in service and minimizing the likelihood of catastrophic failure.

4.1 CAUSES OF FAILURE

In general, service failures may arise because of various frequently inter-related causes. When the initial material selection was made it may have been that one likely cause was clearly identified that defined a principal property, e.g., corrosion resistance of a piece of chemical plant or creep resistance of a gas-turbine engine component, on which the main elements of the design were based. More often than not, however, other factors should also be taken into account. These may range from manufacturing factors, such as formability, to more clearly metallurgical factors, such as fatigue or wear resistance. For most engineering components the causes of failure can be broken down into three groupings:

(i) failures attributable to faulty design considerations or incorrect materials selection
(ii) failures due to faulty processing
(iii) failures due to deterioration during service conditions

Identification of the cause is important since avoidance of failure by *overdesign* through the use of large safety factors is extravagant whereas *underdesign* invariably leads to premature failure. Thus, usually the initial design involves a minimal degree of overdesign and makes use of limited safety factors and other like precautions. It is salutory to examine some statistical data for failure origins, since these clearly indicate where attention should most obviously be concentrated. Table 4.1 presents some data from 350 failure investigations in chemical, mining and manufacturing engineering industries.

Table 4.1 Causes of failure in some engineering industry investigations

Origin	%
Improper materials selection	38
Fabrication defects	15
Faulty heat treatments	15
Mechanical design fault	11
Unforeseen operating conditions	8
Inadequate environment control	6
Improper or lack of inspection and quality control	5
Material mix-up	2

Similar data from 230 laboratory reports on failed aircraft components are shown in Table 4.2.

Table 4.2 Summary of data from laboratory reports on failed aircraft components

Origin	%
Improper maintenance	44
Fabrication defects	17
Design deficiencies	16
Abnormal service damage	10
Defective material	7
Undetermined cause	6

So far we have considered failure without attempting to make a rigorous definition of it. It is perhaps appropriate to do this now since although fracture is often important in failure it is not a *necessary* part of failure.

Failure can be defined as having occurred when a component is *no longer capable of satisfactorily fulfilling its service function* either because of fracture or excessive deformation or deterioration.

The failure *mechanism* is usually a material failure that is controlled by the thermomechanical history of the material during processing and by the service conditions.

Now, the data contained in Tables 4.1 and 4.2 can be re-examined if a subdivision according to the mechanism of failure is made. This is done in Tables 4.3 and 4.4.

Table 4.3 Mechanisms of failures in some engineering industry investigations

Mechanism	%
Corrosion	29
Fatigue	25
Brittle fracture	16
Overload	11
High temperature corrosion	7
Stress corrosion/corrosion fatigue/ hydrogen embrittlement	6
Creep	3
Wear, abrasion and erosion	3

Table 4.4 Mechanisms of failure of aircraft components

Mechanism	%
Fatigue	61
Overload	18
Stress-corrosion	8
Excessive wear	7
Corrosion	3
High temperature oxidation	2
Stress-rupture	1

In identifying causes and mechanisms it is necessary to distinguish between those failures that are largely dependent on design and fabrication and those that are materials dependent. With the former, a slight change in design or production procedure may prevent failure, whereas the selection of a new material for the same design may result in a duplication of a previous failure albeit on a longer time scale. On the other hand there are many instances in which the solution is a materials-based one.

Alongside the three-way division of causes indicated above a second two-way division can be made into:

(i) failures occurring at stresses *above* the design level
(ii) failures occurring at stresses *below* the design level.

4.1.1 Failure at stresses above the design level

In principle these are the easiest to rectify and result either from error, e.g., faulty design considerations or material misapplication, or from overload. In some cases the overload is so unexpected, e.g., in a car crash, that perhaps the designer could be forgiven for not anticipating it. However, this type of overload condition is normally the result of the failure of some other part of the structure for which the designer can be considered responsible. A detailed classification of this group of failure causes with the responsible condition is given in Table 4.5.

Table 4.5 Failures due to faulty design considerations or misapplication of material (after Dolan, 1972)

Failure	Condition
Ductile failure (excess deformation, elastic or plastic; tearing or shear fracture)	Error
Brittle fracture (from flaw or stress raiser of critical size)	Overload*
Fatigue failure (load cycling, strain cycling, thermal cycling, corrosion fatigue, rolling contact fatigue, fretting fatigue)	Error/overload
High-temperature failure (creep, oxidation, local melting, warping)	Error
Static delayed fractures (hydrogen embrittlement, caustic embrittlement, environmentally stimulated slow growth of flaws)	Overload
Excessively severe stress raisers inherent in the design	Overload
Inadequate stress analysis, or impossibility of a rational stress calculation in a complex part	Error
Mistake in designing on basis of static tensile properties, instead of the significant material properties that measure the resistance of the material to each possible failure mode	Error

*Overload usually implies a neglected modification to the design specification which leads to stresses above the design level

With regard to inadequacies of stress analysis referred to in Table 4.5, especially those where the form of loading and component shape are so complex as to obviate accurate analysis, it must be anticipated that prototype testing under simulated service conditions will be seen as an inherent part of the design process.

Furthermore some of the overload conditions will be localized. Thus there will be some overlap in our possible classifications, since some failures at stresses below the design stress must result from local modifications of materials and conditions so that failure is encouraged. Then overload results.

4.1.2 Failure at stresses below the design level

Processing and fabrication procedures are of primary importance in determining the existence of flaws and metallurgical changes that are built into the component structure. Mechanical, thermal and chemical processes can give rise to both microscopic and macroscopic defects, and these may be located either at the surface or in the interior. Commonly the initial defect is itself not catastrophic, but in-service interactions between the environment and/or the applied stresses may lead to defect growth, and a sub-critical defect evolves into a critical defect. In practice it is advisable to assume that all materials are prone to defects and then undertake assessment procedures to evaluate their severity. Some of the defects that occur, particularly those associated with anisotropy of structure and properties such as variations in grain size and shape, may, however, be extremely difficult to identify. Failures produced by defects are most properly considered as very localized overload failures, since it is expected that either the stress conditions locally exceed the design stresses because of stress concentrations or the resistance to stress falls to below the design stress. The causes of failures resulting from faulty processing and fabrication are listed in Table 4.6.

Table 4.6 Causes of failures due to processing and fabrication faults (after Dolan, 1972)

Flaws due to faulty composition (inclusions, embrittling impurities, wrong material)

Defects originating in ingot making and castings (segregation, unsoundness, porosity, pipes, nonmetallic inclusions)

Defects due to working (laps, seams, shatter cracks, hot-short splits, delamination, and excess local plastic deformation)

Irregularities and mistakes due to machining, grinding, or stamping (gouges, burns, tearing, fins, cracks, embrittlement)

Defects due to welding (porosity, undercuts, cracks, residual stress, lack of penetration, underbead cracking, heat-affected zone)

Abnormalities due to heat treating (overheating, burning, quench cracking, grain growth, excessive retained austenite, decarburization, precipitation)

Flaws due to case hardening (intergranular carbides, soft core, wrong heat cycles)

Defects due to surface treatments (cleaning, plating, coating, chemical diffusion, hydrogen embrittlement)

Careless assembly (mismatch of mating parts, entrained dirt or abrasive, residual stress, gouges or injury to parts, and the like)

Parting line failures in forging due to poor transverse properties.

One of the all-too-frequent causes identified in Table 4.6 is careless assembly. No matter what care is taken in design and materials preparation, unless there is the corresponding care in incorporating components into machines and structures, failure will be encouraged. To some considerable extent the designer is in the hands of the fabricators and assemblers, and accordingly attention should be given to assembly inspection in the same way that it is given to component inspection.

The other grouping of causes of failure at stresses below the design level arises because of progressive deterioration in service so that there is either a fall in a critical material property, e.g., as associated with overageing in age-hardened aluminium alloys, or a rise in the effective stress to which the component is subjected. The latter can be caused by removal of material by corrosion or erosion or by the growth of cracks. Deterioration can be caused by mechanical, chemical and thermal effects, and it is extremely difficult to predict performance using standard tests for materials evaluation. Specialized testing is needed that ensures that the critical effect is identified as a controlling variable, e.g., cavitation resistance. Often the specialized testing must also include careful simulation of service conditions. The various causes of failures of this type are given in Table 4.7.

Table 4.7 Causes of failures associated with deterioration during service (after Dolan, 1972)

Wear (erosion, galling, seizing, gouging, cavitation)

Corrosion (including chemical attack, stress corrosion, corrosion fatigue) dezincification, graphitization of cast iron, contamination by atmosphere

Inadequate or misdirected maintenance or improper repair (welding, grinding, punching holes, cold straightening, and so forth)

Disintegration due to chemical attack or attack by liquid metals or platings at elevated temperatures

Radiation damage (sometimes must decontaminate for examination which may destroy vital evidence of cause of failure), varies with time, temperature, environment and dosage

Accidental conditions (abnormal operating temperatures, severe vibration, sonic vibrations, impact or unforeseen collisions, ablation, thermal shock, and so forth)

Apart from the causes listed in Table 4.6 and 4.7 that can be attributed to organisational deficiencies, e.g., careless assembly or misdirected maintenance, and those due to conditions outside those anticipated by the designer and able to be regarded as accidental, the precise identification of the cause of a failure usually requires a clear understanding of the mechanism of failure. This understanding is also of fundamental importance in specifying procedures for the avoidance of failure. Examination of the mechanisms of failure can now appropriately be undertaken as part of a general approach to failure diagnosis.

4.2 DIAGNOSIS OF FAILURE

4.2.1 Mechanisms of failure

The diagnosis of failure requires knowledge of both the microscopic and the macroscopic features associated with the different possible mechanisms of failure. It is important to distinguish between causes and mechanisms. Possible causes have already been outlined. Possible mechanisms, some of which have already been mentioned are:

brittle fracture
ductile fracture
fatigue (at or near the expected life)
low-cycle fatigue
creep
compression failure (buckling)
gross yielding
corrosion

stress corrosion
corrosion fatigue
wear

Mechanisms frequently interact. For instance, wear processes may initiate a crack that then propagates by fatigue. The ultimate failure is a consequence of the entire history of stress, time, temperature and environment. Further, when considering mechanisms of failure it is necessary to know the details of this history with regard to:

   (i) loading mode: whether static, oscillating or impact
  (ii) stress type: whether tension, compression or shear
 (iii) operating temperature: whether low, ambient or high
 (iv) operating environment as it affects corrosion rate

Some failure mechanisms are strongly dependent on time, so a knowledge of the period for which a component has been in service is also essential.

4.2.2 Characteristics of fracture processes

Before the identification of fracture mechanisms can be undertaken, it is important to know the characteristic features associated with the different fracture processes. The broken parts often display markings that constitute a topological map that contains unequivocal information on the sequence of events that preceded ultimate failure.

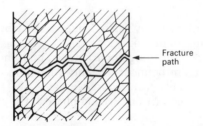

Fracture
path

Fig. 4.2 Schematic representation of intercrystalline fracture path

There is an initial, usually distinct, division based on macrostructural features into *intercrystalline* or *transcrystalline* fractures. In the former, which are the easiest to recognize, the fracture path lies along the grain boundaries of the failed sample (Fig. 4.2). The fracture has a three-dimensional facetted appearance in which the original grain shapes can be clearly distinguished. However, with transcrystalline fractures the fracture path travels across the constituent grains (Fig. 4.3). Then the surface features evident tend to be planar or concoidal and are characteristic only of the failure process and no clear indications of the underlying grain structure are apparent.

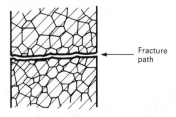

Fig. 4.3 Schematic representation of transcrystalline fracture path

Sub-division can be made within the two initial types of fracture, depending on whether or not the failure is ductile or brittle. Ductile failure may involve either macroscopic deformation, in which there is a substantial change in the sample geometry before failure or only microscopic deformation in which the plastic processes leading to failure are confined to local regions, such as grain boundary zones, in which case there is an absence of macroscopic shape changes.

In the more complex failures, the fracture appearance may be 'mixed mode' involving both intercrystalline and transcrystalline components and ductile and brittle regions. Some consideration will be given to mixed mode failures after an examination has been made of basic failure modes.

Intercrystalline fracture
Brittle intergranular fractures have a classical appearance and principally arise from four causes:

    (i) hydrogen embrittlement, which is characterized by hair-lines and pores at the grain boundaries (Fig. 4.4)
   (ii) segregation of an embrittling impurity, e.g., phosphorous or arsenic in steels, to the grain boundary vicinity (the grain boundary facets are then often virtually featureless: see Fig. 4.5)
  (iii) precipitation of an embrittling second phase at the grain boundaries (Fig. 4.6)
  (iv) stress-corrosion cracking as a result of the combined effects of the applied (or residual) stress and the presence of a corrosive environment.

Intercrystalline stress-corrosion cracks can usually be distinguished from the other types of brittle intercrystalline fracture because the crack breaks up and propagates simultaneously down several boundaries (Fig. 4.7). Further, there are frequently corrosion products on the fracture surface.

In the context of brittle intergranular fracture, it is important to note that with ferrous materials although the sample may be basically ferritic, the failure surface may be associated with prior austenite grain boundaries. Then fracture is consequential upon segregation and/or precipitation occurring in the austenite phase at high temperatures before transformation on cooling.

Fig. 4.4 Intergranular fracture in steel induced by hydrogen embrittlement. 'Hair lines' on the grain boundary facets can be clearly seen

The most prominent feature of ductile intercrystalline failures is the presence of fine dimples on the grain boundary facets caused by the ductile growth and linking together of microcavities. These microcavities can be of two distinct origins, either:

   (i) due to high temperature creep and creep rupture processes occurring at temperatures $\geqslant 0.5 T_m$, where T_m is the absolute melting temperature, or

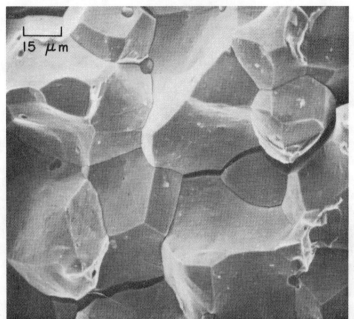

4.5

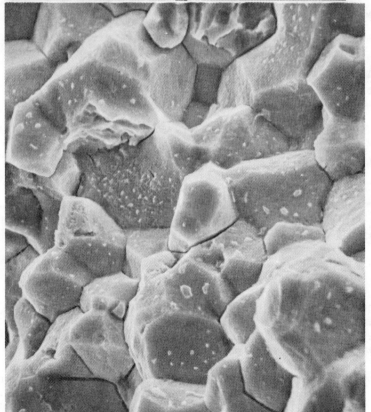

4.6

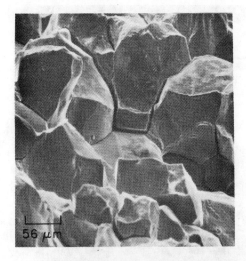

Fig. 4.7 (a) SEM fractograph of intercrystalline stress corrosion fracture. (b) Micrograph of a section through fracture showing multiple crack propagation: magnification 100×

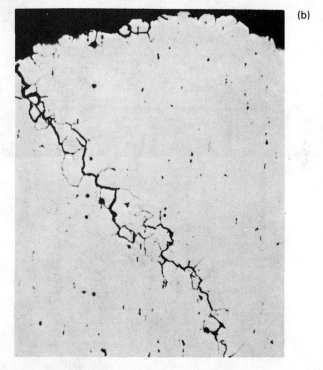

Fig. 4.5 Intercrystalline failure in Armco iron resulting from grain boundary segregation (SEM fractograph)

Fig. 4.6 Grain boundary precipitates of TiC in a martensitic steel produced this intercrystalline fracture: magnification, 2000×

(a)

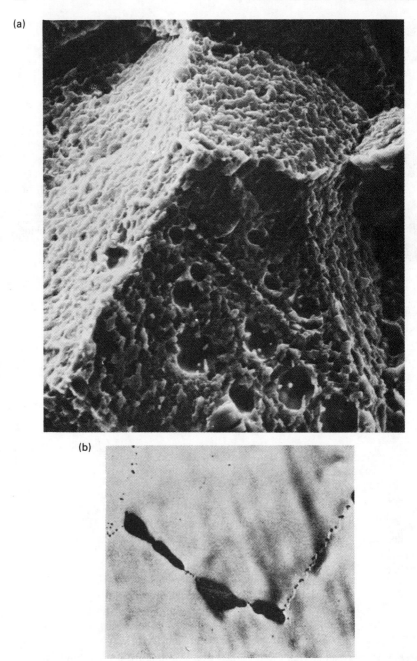

(b)

Fig. 4.8 (a) Creep voids on grain boundary fracture surface of a nickel alloy tested at 800°C: magnification 1340×. (b) Grain boundary voids near a creep fracture surface: magnification 100×

(ii) because of plastic slip processes in the grain boundary zone at temperatures $\geqslant 0.5T_m$, in which microvoids grow in a manner similar to that of macroscopic void growth in classical ductile fracture (see below).

These two types of ductile intergranular fracture (Figs 4.8a and 4.9a) are often difficult to distinguish since the ultimate failure surfaces are very similar. However with the higher temperature failures there is usually evidence of void nucleation and growth on transverse grain boundaries away from the fracture surface associated with the accommodation of creep-induced deformation (Fig. 4.8b). The lower temperature failures normally require a soft zone, such as a precipitate-free zone in an age-hardening alloy, to which the plastic deformation can be confined.

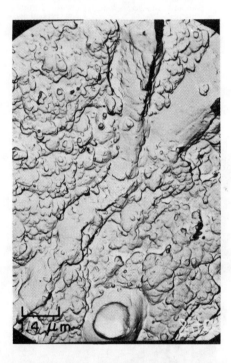

Fig. 4.9 Microductile grain boundary dimples in the precipitate free zone of an aluminium alloy which failed by intergranular fracture: magnification, 7000×

The characteristics of intercrystalline failures which can be made use of in diagnosing failure mechanisms are summarized in Fig. 4.10. In some cases there may be more than one contributory cause, in which case the fracture will exhibit multiple features, e.g., a combination of those in figs. 4.4 and 4.9 as shown in Fig. 4.11.

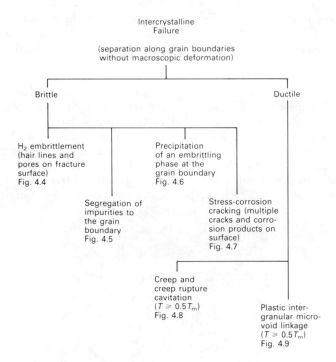

Fig. 4.10 Summary of intercrystalline fracture mechanisms

Fig. 4.11 Mixed mode intergranular brittle/ductile failure in a steel

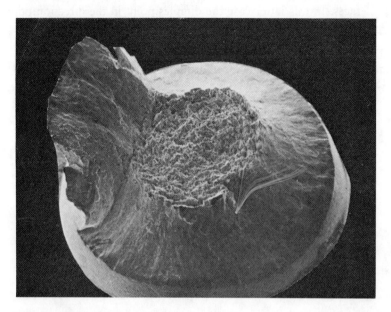

Fig. 4.12 Ductile cup-and-cone fracture in a tensile test piece showing a well developed shear lip: magnification, 18×

Transcrystalline fracture

Transcrystalline fractures can be classified as ductile, microductile or brittle.

Ductile failures involve significant amounts of plastic deformation and fracture occurs by the growth during plastic deformation of internal voids that join together to produce complete separation. The fracture surface has a fibrous appearance and often shear lips occur in the zone of final fracture (Fig. 4.12). When this is the case the origin of the failure can frequently be identified because of its remoteness from the shear lips. Ductile fracture of this type involves slow crack growth and the cross-section at fracture may be reduced by necking associated with plastic instability. The fracture surface is covered with dimples formed as the growing voids link up and their geometry can be made use of to identify the mode of crack growth. Three characteristic modes can be distinguished:

   (i) a normal tensile growth mode in which rounded dimples occur on both fractures faces
  (ii) a shear mode that leads to elongated dimples pointing in the direction of shear, i.e., they point in *opposite* directions on the matching fracture faces, and
 (iii) a tearing growth mode that also leads to elongated dimples, but in this case they point in the *same* direction on the matching fracture

faces. In this latter case the dimples point in the direction opposite to that of crack growth.

These different modes are illustrated schematically in Fig. 4.13 and examples of normal and elongated dimples are shown in Figs. 4.14 and 4.15, respectively.

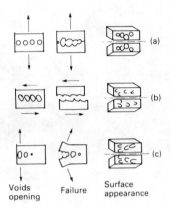

Fig. 4.13 Various forms of ductile fracture showing characteristics of dimple formation: (a) tensile; (b) shear; (c) tearing

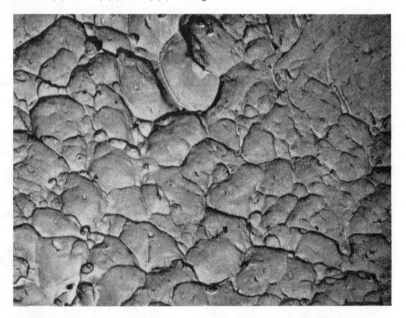

Fig. 4.14 Equiaxed dimples formed during ductile tensile failure of steel: magnification, 8000×

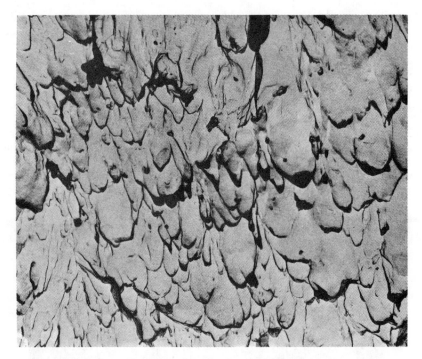

Fig. 4.15 Elongated dimples formed during ductile shear failure of steel (X10000): magnification, 10 000×

There is not normally a significant amount of macroscopic plastic deformation associated with microductile intercrystalline fracture, the most classical form of which is fatigue failure. The fracture surface is commonly flat and normal to the principal stress axis. After fatigue failure, prominent surface markings known as 'beach marks' or 'shell markings' can often be clearly seen on the fracture surface (Fig. 4.16) and provide important information about the site of fracture initiation. Fatigue occurs under cyclic load conditions, and on a microscopic scale fatigue cracks propagate discontinuously in response to the cyclic nature of the imposed stress. This gives rise to striations on the fracture surface, which are the most distinguishing feature of fatigue failures (Fig. 4.17). Fatigue crack propagation continues until the remaining cross-sectional area can no longer support the applied load, at which stage fast fracture occurs. Fatigue striations arise from the successive opening and closing of the crack under load, and although their width and form varies depending on the material and the loading conditions, they can provide useful information on behaviour during the more critical stages of crack growth. However, fatigue striations do not occur during the very early stages of crack formation, the so-called 'incubation' stage. This represents an important proportion of the life of the sample, and any factors that contribute to the enhancement of crack initiation will reduce the life accordingly. This is especially true of conditions that induce surface damage. For example, the various forms of surface wear that can occur namely, adhesive wear (also known as 'galling' and shown in Fig. 4.18), abrasive wear (Fig. 4.19) and fretting (corrosion-assisted wear, as illustrated in Fig. 4.20) are themselves types of microductile failure that can give rise to subsequent fatigue failures. In practice, when fatigue loading occurs in association with corrosive conditions enhanced rates of crack growth result and a corrosion-fatigue failure is caused. Corrosion-fatigue failures differ from conventional fatigue failures in that multiple crack growth is common as can be seen in Fig. 4.21. Here also, as with intergranular stress-corrosion cracking, it is usual to be able to identify corrosion products on the fracture surface.

Fig. 4.16 Macroscopic view of a fatigue failure of a steel shaft showing well-developed beach marks. The point of crack initiation is arrowed

Fig. 4.17 Fatigue striations: (a) in a nickel alloy tested at 650°C: magnification, 550×; (b) in an aluminium alloy (fractured intermetallic particles can also be seen): magnification, 1500×

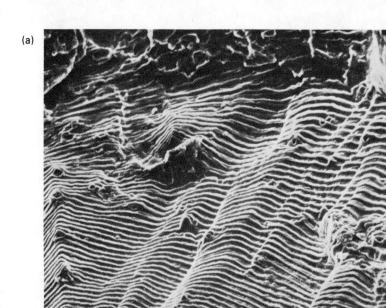

(a)

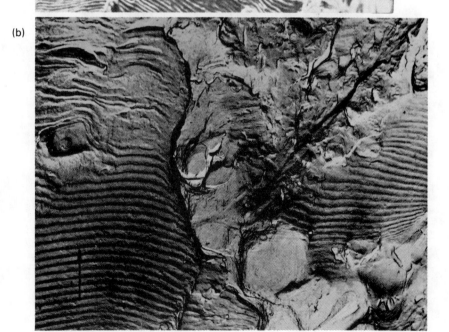

(b)

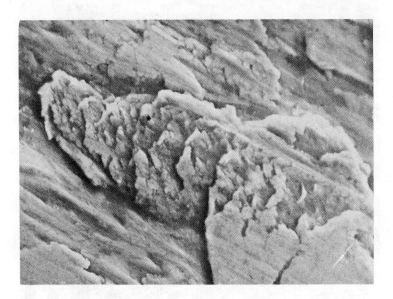

Fig. 4.18 Adhesion marks on a gear tooth produced by momentary friction welding: magnification, 1100×

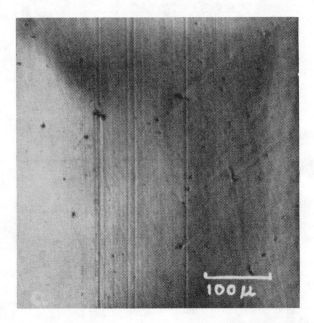

Fig. 4.19 Ploughing marks formed during sliding abrasion of steel

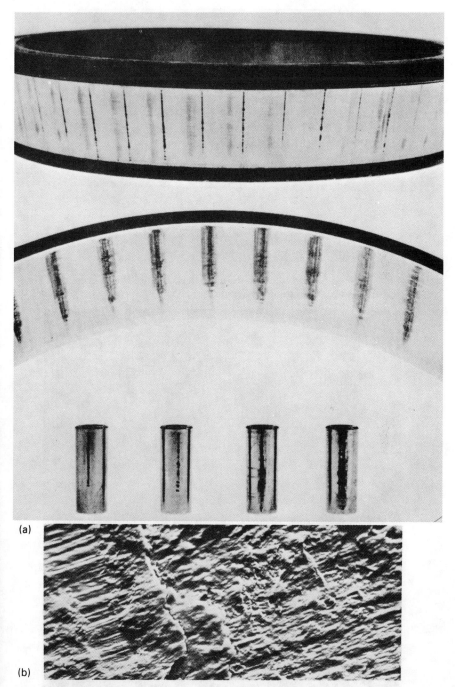

Fig. 4.20 (a) Fretting of cone, cup and rollers in a steel taper roller bearing caused by vibration, (b) Detail of fretted surface showing characteristic terracing: magnification, 180×

Fig. 4.21 Corrosion fatigue cracks in stainless steel. Multiple branched cracks are characteristic of corrosion fatigue: magnification, 75×

Fig. 4.22 A cleavage facet showing river lines formed during brittle fracture of a high chromium steel: magnification, 650×

Fig. 4.23 Chevron marks on a fractured steel plate. The fracture originated on the right and the chevron marks point towards the initiation site: magnification, 5×

Brittle transcrystalline fractures occur with little or no prior plastic deformation. Crack propagation is by cleavage, which leads to planar facetted fracture surfaces with microscopic 'river lines' on the grain facets (Fig. 4.22). The facetted nature of the fracture surface (usually with one facet per grain) can give the specimen a reflective crystalline appearance. Brittle cracks propagate extremely rapidly and often there is noise emission at the same time. On a macroscopic level large specimens which have failed in a brittle mode frequently exhibit chevron markings, as can be seen in Fig. 4.23, which point back to the point of fracture initiation. Study of chevron markings can sometimes reveal that failure has been associated with simultaneous, or almost simultaneous, multiple crack initiation.

A special form of transcrystalline brittle failure is that associated with crystallographic stress corrosion. This produces a form of quasi-cleavage with many of the features of normal cleavage even though the mechanisms of crack propagation are quite different. However, the surface markings

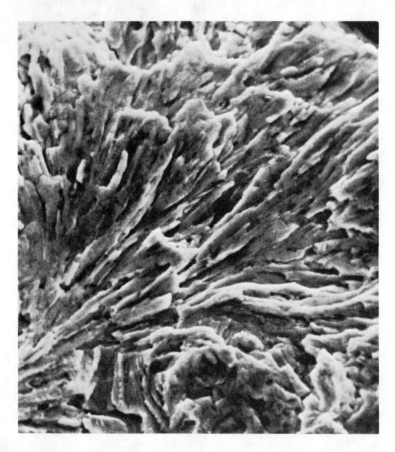

Fig. 4.24 Featherlike transcrystalline fracture surface of an austenitic stainless steel that failed by stress corrosion: magnification, 2000×

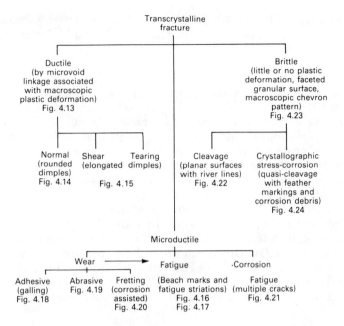

Fig. 4.25 Summary of transcrystalline fracture mechanisms

differ from river lines being more feather-like (see Fig. 4.24) and there is usually corrosion debris on the fracture surface.

Fig. 4.25 is the corresponding figure to Fig. 4.10 for intercrystalline failure, and summarizes the characteristics of transcrystalline failures.

4.2.3 The methodology of failure analysis

With knowledge of the mechanisms of fracture and the characteristics of fracture processes it is now feasible to define a systematic approach to identification of the causes of fracture. Armed with this assessment of performance in service, the engineer can draw conclusions about the suitability of the material(s) used and proceed to recommend any design modifications or alterations in processing routes or materials specifications. The basic steps to be followed may not all pertain in every instance, and on occasions there may be slight variations in the order of precedence. Nevertheless there are four areas of operation. When giving attention to these areas it is important to ensure that the nature and sequence of the examination is such that evidence required at later stages is not contaminated or destroyed at earlier stages. The four areas are:

   (i) initial observations
   (ii) assembling background data
   (iii) laboratory assessments
   (iv) analysis of data and synthesis of the failure

This last stage will necessarily also involve report preparation as appropriate, incorporating recommendations. Let us consider these operations in turn.

Initial observation

As soon as possible after the report of a failure it is essential a visual inspection (where practicable), of the component and the assembly of which it was a part should be made, preferably at the site. In the meantime, instructions should be issued to ensure that (as far as possible) all parts are protected from further damage or deterioration, say by corrosion. In making the visual examination it is appropriate to prepare a comprehensive photographic record (in colour if this is feasible). Detailed interpretation must be made of macroscopic deformations, presence of stress concentration or contributing imperfections, final location of components and debris, sizes and other physical data, fracture appearance, deterioration, contamination and other factors. At this stage it should be possible to draw preliminary conclusions about the predominant failure mode and mechanisms and the direction of crack propagation and sequence of failure, which may involve more than one component. Attention can then be concentrated on acquiring background data.

Background data

In this stage it is necessary to gather together all available data concerned with relevant codes of practice, specifications and drawings, particularly those data relating to the critical component design. In addition, details of the fabrication and heat treatment procedures adopted, any repairs, maintenance schedules and service loading (as far as it is known) should be obtained. With regard to load conditions, not only normal service loads (both static and dynamic) should be ascertained, but also any information available should be gained concerning such factors as accidental overloads, adverse environmental conditions or defective lubrication. In this latter context both corrosive conditions and thermal conditions are of importance. Whenever possible these data should be supplemented by quality control, servicing and inspection reports prepared both before and during service.

Unobvious factors should not be overlooked, e.g. conditions of materials handling, storage and *identification* should be ascertained.

Laboratory assessments

The above data can then be supplemented with the results of tests to verify the properties the component under examination actually possesses. Ideally these tests should largely be non-destructive, perhaps embracing radiography, ultrasonic testing, eddy-current testing, etc. (see Chapter 6), as well as including more conventional assessments such as hardness testing, *in situ* chemical analysis. The aim is to determine the extent to which the component has the required composition, dimensions and structure. Then attention can be concentrated on fractography (using scanning microscopy) and metallography. The latter is necessarily destructive, and care is needed to ensure that no important evidence is destroyed. In many critical cases, the newer facilities that allow high-resolution *in situ* chemical analysis can be used to advantage. When performing fractographic and ancillary surface examinations it is advisable to bear in mind the attributes and characteristics given in Figs. 4.10 and 4.25, since they should lead to a tentative identification of the mechanism of failure and the site of failure initiation.

Synthesis of failure

With all the above to hand, a list can be prepared of both positive and negative facts and evidence. Sometimes it is important to know that specific things did not happen or certain evidence did not appear to determine what could have happened. From a tabulation of these data, the actual failure should be able to be synthesized to include all available items of the evidence. Frequently, it may be necessary to develop additional data for completeness or for further verification. This could involve simulation testing under the conditions prevailing during manufacture or in-service. Where accelerated test data must be used, e.g., in dealing with creep failures, these data should be treated with caution.

To ascertain the true cause of a failure, the engineer/metallurgist must give full consideration to the interplay of design, fabrication, materials properties, environment and service loads. The cause will usually be classified in one of the categories outlined in Tables 4.5, 4.6 and 4.7, and corrective action or applied research guidance can be recommended. As indicated previously, appropriate solutions may involve redesign, change of alloy and/or processing, quality control, protection against environment, changes in maintenance schedules, or restrictions on service loads or service life.

Follow-up on the recommendations is frequently a difficult task, but should be undertaken for the more critical failures. Cooperation between the investigator, the designer, the manufacturer and the user is crucial in developing good, workable changes. Through using the iterative procedures indicated in Fig. 4.1 based on a knowledge of performance in service, the engineer and the metallurgist can work together to improve fitness for purpose and to ensure that the properties considered to be of importance in engineering design (see Chapter 3) are used to optimum advantage.

4.3 CASE STUDIES

The use of case studies and reference atlases of damage and fractography is a valuable and effective way of gaining insight into the complexities and vagaries of performance in service. A series of most useful texts and sources books is listed at the end of the chapter. A typical case study embodies all the factors described in Section 4.2.3, giving appropriate procedural and factual details. Thus a cumulative view can be built up based on sound experience. There are several classic case reports that are readily available in full or abridged versions. Many of the latter are contained in the source books listed at the end of the chapter. A typical selection from these would comprise the following very varied list:

Fatigue and the Comet disasters
Failures of forged end bells on large electric generators
Corrosion-fatigue failures of boiler tubes
Flame ring failures at weldments
Deterioration of brass condenser components by impingement
Forging defects in aqualung cylinders
Galling of stop valve spindles
Failure of a 95–5 copper–nickel–iron line by overheating
Aircraft hydraulic line failure
Wrought iron link failure by impact loading
Aluminium scaffold clamp failure by stress-corrosion cracking
Waveguide flanges–a casting problem
Monel shaft failure by intergranular attack
Heat treating quench cracks in a crosshead pin
Fatigue failures of bus steering assemblies
Diesel crankshaft failure
Thermal cracking of cylinder heads
Corrosion failure of ball bearings
Failure of non-shot-peened springs
Hydrogen cracking in low alloy steel shafting
Fracture of nitrided piston rod
Creep fracture of tubes in steam generating plant
Drum wall destroyed by sulphur attack
Scaling of resistance elements
Grinding cracks

Details of case studies are beyond the scope of the present work, but the use of case studies is strongly recommended. In many instances when considering a failure examination it will be found that there already exists a parallel case study report in the literature.

SUGGESTIONS FOR FURTHER READING

Source Book in Failure Analysis American Society for Metals, Ohio (1974).
R.D. Barer and B.F. Peters (1970) *Why Metals Fail*, Gordon and Breach, New York.
V.S. Colangelo and F.A. Heiser (1974) *Analysis of Metallurgical Failures*, John Wiley, New York.
T.J. Dolan (1972) Analyzing failures of metallic components, *Metals Eng. Q.* 12(4), 32–40.
L. Engel and H. Klingele (1981) *An Atlas of Metal Damage*, Wolfe Publishing, London.

J.L. McCall and P.M. French (eds) (1978) *Metallography in Failure Analysis*, Plenum Press, New York.

Metals Handbook. Vol. 9, Fractography and Atlas of Fractographs (1974); Vol. 10, Failure Analysis and Prevention (1975) American Society for Metals, Ohio.

CHAPTER 5

Quality assurance

5.1 INTRODUCTION

Component failure is serious, not only in the financial loss that may result, which may possibly be covered by insurance, but also in the uninsurable losses that follow, such as loss of good name, good will or even the imposition of a prison sentence. Clearly the quality of a product must be maintained at a level that is not only economically acceptable, but also ensures that consequential losses, if failure were to occur, are minimized.

Quality assurance is the name given to a manufacturing procedure whose object is to ensure conformance of the manufactured parts to the requirements specified by the design and materials' engineers. Quality itself may be defined as a degree of excellence, and the chosen level of quality of design for a manufactured part or assembly is usually a management decision that balances the cost of ensuring that level of quality against the cost of possible failure. The design and materials' engineers translate that decision into manufacturing and inspection requirements.

The quality engineer is concerned that the quality level specified can be achieved in a consistent manner throughout the manufacturing life of a component. The minimum degree of continuous conformance to require-ments must be the quality level laid down in drawings, instructions and other documentation.

The word assurance relates to a guarantee, and therefore quality assurance is concerned with all activities and functions dealing with the attainment and maintenance of production quality.

Quality cannot be inspected into any material or part after manufacture. It follows that inspection of the parts at the final stage in production is an 'after the event' situation that will not prevent defective work being produced. Inspection is only a part, though an essential part, of the overall concept.

 Quality assurance covers all activities and functions concerned with the attainment of quality, and within this definition quality control may be interpreted as a system for programming and co-ordinating the efforts of various groups in an organization to maintain or improve quality at an economical cost.

5.2 QUALITY MANUAL AND QUALITY PLAN

The Health and Safety at Work Act (1974) lays down that every designer and manufacturer is liable for the safety of their products, and should take reasonable steps to ensure that safety. Evidence in the form of a *quality manual*, detailing all the steps taken to ensure that safety and satisfactory performance of a component or assembled unit, is an effective basis for a defence against legal actions alleging failure of a manufacturer and his design and production engineers to meet their responsibilities on safety and performance.

A quality manual is the keystone of a quality assurance system. It is a document setting out the general quality policies and practices of an entire organization. A *quality plan* is a document setting out the specific quality practices and activities relevant to a particular project or contract.

5.3 QUALITY ASSURANCE IN PRACTICE

Quality assurance procedures commence at an early stage in the production of a metal or component. Thus, in a metalliferous mining and refining company, the quality control engineer will be called upon to confirm the metallic content and suitability for processing of the ore deposit and to check characteristics of the ore, such as the particle size to which the ore must be crushed and ground, the processes required to separate the mineral from the gangue and the metal from the mineral to achieve efficient extraction of the metal from the ore. The composition of an iron ore, for example, varies as the deposit is mined and just one of the requirements of the quality assurance procedure will be the periodic analyses needed to check the iron and silica contents and the presence of any other elements, such as phosphorus, that might affect adversely the quality of the iron to be produced and therefore its market value.

Iron ore is smelted sometimes mixed with scrap iron or steel to yield pig-iron that will normally be purchased to the requirements of a material specification. This specification may include a clause requiring that 100% iron ore and no scrap metal may be used in the production of the pig-iron. This restriction controls the chemical composition and reduces the chance of contamination by undesirable elements. Such material is usually accompanied by a certificate of analysis and load identification.

In the production of iron castings, pig-iron is no longer the only basic raw material, since steel scrap is also used in modern foundries. Not all types of steel scrap are acceptable, however. Free-cutting steels, spring and tool steels, gear wheels and rail points, for example, are unacceptable because they contain alloying elements that would have damaging effects on the properties of the castings, and the quality control engineer would have the responsibility for guarding against the use of such unacceptable material.

Thus, in this example of the production of an iron casting, the quality manual would specify the raw materials to be used, possibly the source of those materials, the analyses to be carried out on the raw materials and the finished product; it would lay down the melting and casting procedures, the procedure and the materials to be used in the preparation of the mould and any other aspects of the production of the casting that would effect its quality.

5.4 MONITORING MACHINABILITY

Advances in technology have generated a series of low-carbon steels of consistent quality, improved machinability, and capable of giving a better surface finish. Such materials normally have large spherical sulphide deposits between the dendrites. They also contain lead, which substantially increases the machinability over that attributable to sulphur alone. The lead provides additional lubrication at the tool-tip, which reduces the tool's working temperature and hence tool-wear.

The influence of these elements and others, such as the ferrite-embrittling elements phosphorus and nitrogen, and the deoxidants aluminium and silicon, is shown in Fig. 5.1. Sulphur in the form of manganese sulphide particles has the most significant influence on machinability. Ideally, these particles should be thick oval shape and be evenly distributed in the finished bar. The ideal formation is determined by a high level of oxidation in the melt, which requires that such elements as silicon and aluminium must be kept to a minimum. This requires strict control of the steel-making process which also controls the phosphorus and nitrogen content. Figure 5.2 shows the influence of manganese sulphide on machinability, and Fig. 5.3 shows the effect of increasing silicon content.

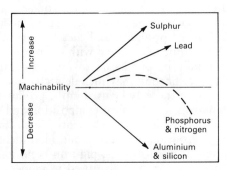

Fig. 5.1 Influence of various elements on machinability

The determining factors in achieving a high production rate from machines using cutting tools are tool-wear, cutting force, power requirements and surface finish, although tool-wear is the most frequent reason why cutting machines need resetting. In a modern machine shop, using computer-controlled automatic tools, a consistent quality of material must

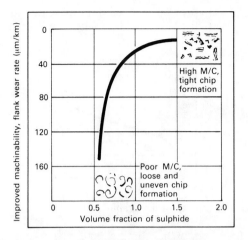

Fig. 5.2 Influence of manganese sulphide on machinability

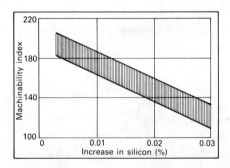

Fig. 5.3 Influence of an increasing silicon content on machinability

be available and the production engineer must call on the materials engineer and the quality organization to specify the requirements in detail, again, even to the raw materials and melting procedure to be used in the production of that material. By using quality control charts to record results of sampling and measurement, and by reference to machinability rating graphs, a patrol inspector can predict the need for a tool-change before defective parts are actually produced (see Figs. 5.4 and 5.5)

An alternative to patrol inspection is *process quality control* — that is, providing the machine operator with suitable means to check the work himself. He will do this at some agreed frequency say, every fifth, tenth, or twentieth component. Provided the check indicates that dimensions are being maintained within certain limits, the checks continue at the prescribed frequency. When the dimensions stray into the warning area (marked on a measuring gauge) the next consecutive item must be checked. If that one also is in the warning band, the machine must be

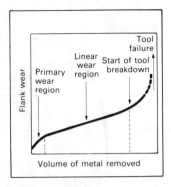

Fig. 5.4 A typical machinability rating graph

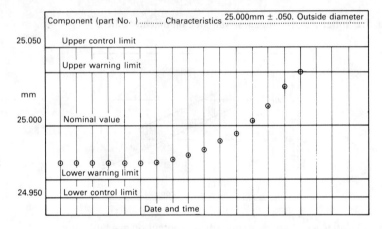

Fig. 5.5 Patrol inspection chart from which tendencies to drift can be detected

reset, and the setter uses the operator's gauge to verify that the newly set machine is producing satisfactory parts.

5.5 INSPECTION PROCEDURE CARDS

Raw materials are normally obtained from approved suppliers to an approved specification which is clear and unambiguous. The type of goods inwards examination is specified on an inspection procedure card (Fig. 5.6), which quotes the characteristics to be inspected, the method of check, the individual number of the gauge to be used and the acceptable quality level (AQL), which may be expressed in percentage defective, or number of defects per 100 items. This is linked to the procedure for altering the

COMPANY LIMITED

GOODS INWARDS INSPECTION
INSPECTION PROCEDURE CARD

Document reference
Page

Sheet   of

Ref. No:
Size No:
Description
Date

Pt. No:
Issue:
Colour:

	A	B	C	D	E	F	G	H	I	J	K	L	M	N	O	P	Q	R	S	T	U	V

Quality characteristics — Zero – 100% defect free

Method of check

			T	N	R		T	R	A	REJ AT
0.4 AQL	up to 500		50	32	13			5		1
1.0 AQL	501 to 3,200		200	125	50			20		2
AUDIT	3,201 to 35,000		200	200	80			32		3
GAUGE No.										AQL

Fig. 5.6 Section of inspection record card. For further details concerning sampling procedures reference should be made to BS 6000, BS 6001 and BS 6002

sampling plan to become more strict when the process average is worse than the specified AQL, and to reduce the amount of inspection when the process average is better than the AQL.

The initial level of inspection would be 'normal', but when all preceding 10 lots have been accepted then 'reduced level' may be introduced. When 'reduced' inspection is in force, 'normal' inspection must be re-introduced if a lot is rejected or the regularity of production is affected in some way. 'Tightened' inspection is specified if 2 lots out of 5 consecutive lots have been rejected at the 'normal' level, and if 10 consecutive lots remain on the 'tightened' level action to improve the quality of the incoming material is usually necessary.

When accepted by goods inwards inspection, the parts are passed to the bonded stores as an identified batch for subsequent issue to production department.

5.6 DEFECTS AND MODIFICATION TO QUALITY MANUAL

Despite careful goods inwards inspection, problems can occur. As example of such a case is a failure that was caused by liquid metal embrittlement. Practically all steels, and some other metals, are susceptible to penetration embrittlement if stressed in tension applied or internal and residual from prior thermal and mechanical histories while in contact with certain molten metals such as copper, zinc, tin and lead. It can occur without evidence of earlier penetration by the liquid metal at rates greater than can be accounted for by diffusion processes, particularly at high temperatures.

While machining a drive shaft, the machine operator inadvertently dropped the shaft on the floor, where it broke into two pieces. As steel forgings can normally be thrown down with impunity, the operator reported the matter to his foreman who, in turn, brought in the section inspector and finally the quality manager. The break showed a crystalline fracture and the parts were referred to the metallurgical laboratory for opinion. In due course, the designer heard about the problem and correctly referred it to the metallurgist.

From their experience, the metallurgical laboratory staff knew that the part was produced by *upset forging* (the enlargement of a part from a smaller to a larger section), the forging temperature being developed by electrical resistance heating. One electrical contact was provided by a split circular 'grip' and applied to one end of the bar to be forged; the other end of the bar butted against a second metallic contact built into the heater section of the forging machine. The heated bar was then transferred to a die and upset in the usual way.

Suspecting the cause of the brittleness, from their knowledge of the manufacturing process, the metallurgical staff had a metallographical examination carried out, which revealed areas of intergranular attack and penetration by fine fingers of copper.

A visit to the supplier of the forgings confirmed that the location of the fracture coincided with the position of the copper grip. Internal arcing between the grip and the bar, because of poor electrical contact, had transferred molten copper to the ground, oxide-free surface, which itself was rapidly heating. The rolled bar retained a fair degree of residual stress from the rolling process and the presence of such stress is a condition for liquid-metal penetration.

Having established the cause of fracture and finding that this was a special case outside normal inspection coverage, the next step was to install a suitable inspection system to determine whether to scrap the entire consignment or even to call in, and scrap, all the parts that had already been manufactured. After assessment of the situation, it was decided that all rough forgings and those partially machined before heat-treatment were to be mechanically distorted over the area of the heating grip by bending them over a set radius and then straightening them again. Embrittled forgings would fail and the tension–compression system is reversed on straightening, so that no affected surface could escape. For those parts tested by mechanical distortion, the effects of such treatment would be eliminated at a later heat-treatment stage. The grip area was to be examined with particular care for any evidence of cracking. Parts already heat-treated were to be examined in the laboratory by etching, binocular examination and fracture testing, to establish whether it would be possible to sort out any sound components from the batch.

5.7 ROUTINE TESTING AND SAMPLING

A quality assurance manual will specify the tests to which a part or assembly must be subjected during or after fabrication or assembly, to determine suitability for further processing. The tests to be used will be defined with acceptance or rejection criteria. The test procedure will be stated with how many items are to be tested and how they shall be selected.

The sample size, that is the percentage of parts to be tested, will be laid down, based on the criticality of the item, the experience with the process involved, the nature of the test and its cost and previous service experience with the part.

Typically, one of the tests most commonly performed on metal components is the hardness test. The hardness tests in common use measure the resistance of a material to permanent deformation by indentation by a Brinell ball, a Rockwell diamond cone, a Vickers pyramidal cone or some other device. A hardness test will determine whether a heat treatment has been completed satisfactorily, and there is a degree of relationship between hardness, tensile properties, fatigue properties and the like.

Hardness may also be measured indirectly. Thus small parts produced in large numbers — fasteners, for example — are now hardness-tested after heat-treatment continuously and automatically using hopper-fed electro-magnetic systems that direct satisfactory components to the appropriate container, and unsatisfactory components elsewhere for correction.

Having unified the structures by heat-treatment, the hardness developed is directly related to the metallurgical phase or phases present. A measurement of the electromagnetic characteristics of the phase is, therefore, relatable to the hardness.

Inspection of 100% of a batch of parts or assemblies is frequently neither practical nor economical, but sampling involves risks that faulty items are overlooked. Many quality engineers use sampling plans obtainable from published tables and these are based on probability theory.

Descriptions of other non-destructive tests available to determine the quality of a product are given later in this chapter.

5.8 CALIBRATION AND MAINTENANCE OF TEST EQUIPMENT

The calibration and maintenace of all metrological equipment is a fundamental requirement for effective quality control. Balances and other weighing instruments, standards for measuring temperature, pressure, volume, viscosity, electrical phenomena and optics may need specialized attention, and testing machines for determining stress, strain, ductility, impact, hardness and fatigue require re-calibration by equipment manufacturers or other specialists.

All test equipment and standards should be supported by certificates, reports, or data sheets confirming the date, accuracy and conditions under which calibration was undertaken, and labelled to show the name of the person who last calibrated the instrument, the date of the last calibration and the scheduled date for the next. Access to devices that can be adjusted but which are fixed at the time of calibration needs to be safeguarded to prevent alteration by unauthorized persons.

In a typical case, a tensile testing machine would be installed and calibrated by the manufacturer and certificated using certified load standards. Adjustment points would be guarded by seals that would break and remain broken if they were tampered with. Re-calibration of the equipment would be required every twelve months and adhesive labels would be attached to inform laboratory staff and visiting quality assurance staff about the status of the instrument.

5.9 RELIABILITY TESTING

Apart from service records, reliability testing by the manufacturer is the most accepted method of proving that the desired standards of quality and fitness for purpose have been achieved. Reliability testing is frequently a joint engineering, metallurgical and quality control exercise using one set of equipment for the common purpose of confirming design satisfaction, proving materials and establishing acceptable manufacturing levels. Testing may be carried out using several 'in-house' testing machines that operate by stressing components or assemblies in exactly the same way as would occur in service.

Inspection teams need target quality levels, good inspection equipment, and the correct ambient conditions to carry out their work properly. As a rule, all characteristics to be examined should be measurable, otherwise they may become a matter of opinion. If the characteristics are not measurable, the provision of actual component standards is a useful expedient. Inspectors work to the standards they are given, so it is important that they are instructed with precision and in writing. Zero defects may be the target standard but, in practice, conformance to the drawing and specified requirements is the quality level accepted.

5.10 MODERN TRENDS IN QUALITY ASSURANCE

The cost of quality is the expense of doing things wrongly and the target must be to reduce the amount of quality control required. Naturally, the use of more highly developed engineering designs will demand higher quality standards, more extensive training and the introduction of completely automatic inspection systems. As described earlier, hardness measurement can be carried out using electronic evaluation of forward and rebound velocities convertible to Brinell, Vickers or Rockwell scales, coupled to recording devices with audible alarm for non-conformance to upper and lower limits of acceptability. It is also possible to measure surface texture using a microprocessor with solid-state memory, touch-button switch to other stored programs, visual display unit and operator reminder. Micro-surgical techniques have been borrowed for magnification and illumination. Closed-circuit television is now being installed on inspection microscopes, and highly accurate height and depth measuring instruments with remote digital read-out and co-ordinate measuring machines fitted with calculators, computer assistance and thermal printers for hard-copy results are already available.

SUGGESTIONS FOR FURTHER READING

British Standards Institution. BS 6000, *Guide to the use of BS 6001: BS 6001, Sampling procedures and tables for inspection by attributes* (Identical in technical content with Defence Specification DEF-131-A); BS 6002, *Specification for sampling procedures and charts for inspection by variables for percent defective.* BSI, London; HMSO, London (for Defence Standards).

The New Quality Requirements for Defence Procurement (1970) Defence Quality Assurance Board Executive, London.

W. Rostoker, J.M. McCaughey, and H. Markus (1960) Embrittlement by Liquid Metals, Reinhold, New York and Chapman & Hall, London.

British Steel Corporation, Carbon Steel Works Group, Rotherham.

British Calibration Service, National Physical Laboratory, Teddington, Middlesex.

CHAPTER 6

Non-destructive testing

6.1 INTRODUCTION

As the name implies, non-dstructive testing (NDT) is simply testing without destroying the component, product or structure. It is an attempt to ensure that components are defect free and 100% inspection can be carried out if justified.

It is not always easy to justify 100% inspection in economic terms, since failure may be little more than a nuisance, e.g., a car door handle that snaps off in service. This is unlikely to do more than annoy the customer and perhaps tarnish the image of the car maker. If, however, a component in the steering system fails, it is a much more serious matter and may result in costly litigation or even the recall of vehicles. In this case, the cost to the manufacturer of not using NDT and 100% inspection can be very high, and NDT is clearly justified. The 'Health and Safety at Work' Act and legislation on product liability will considerably increase the costs of failures and breakdowns to the manufacturer and hence will increase the need for NDT. To meet financial libailities, the alternative is to take out costly insurance but there is no insurance against loss of a good name nor against possible imprisonment of those held responsible.

Thus NDT is becoming commonplace in the manufacture of safety critical components and structures and is likely to become cost effective for many less critical components in the future.

6.2 DEFECTS IN COMPONENTS AND STRUCTURES

During the manufacture of components and structures, there are many different mechanisms by which defects can be produced:

  (i) defects produced by shrinkage in castings and welds (Fig. 6.1)
 (ii) defects produced by non-uniform or excessive deformation in wrought products (Fig. 6.2)
(iii) defects produced by thermal stresses in the manufacturing cycle, e.g., grinding cracks (Fig. 6.3)
 (iv) defects produced by transformation stresses (martensite formation), e.g., quench cracks (Fig. 6.4)

166

(v) inclusions
(vi) gaseous defects, such as blow holes and hydrogen cracks

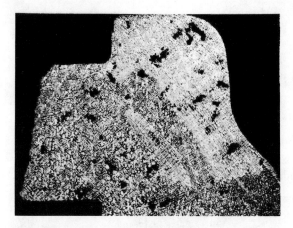

Fig. 6.1 Micro-section of shrinkage porosity in small casting

Fig. 6.2 V-shaped 'cupping' defect revealed in a bolt

After components have been in service new defects may be produced by the service conditions. The most common of these are:

(i) Fatigue and corrosion fatigue cracks
(ii) Stress corrosion cracks (Fig. 6.5)
(iii) Creep cracks

It is important to understand the origin and nature of defects in order to devise suitable non-destructive inspection techniques. In any chapter on NDT it is essential to discuss the types of defects that can occur in various products and these will be described in the next section.

Fig. 6.3 Grinding cracks revealed by dye penetrant inspection

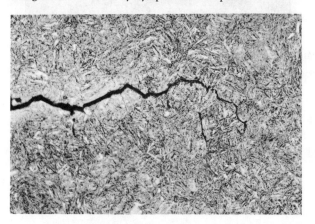

Fig. 6.4 Quench cracks in hardened steel as revealed by microexamination

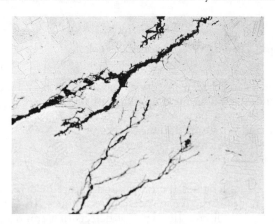

Fig. 6.5 Stress-corrosion cracks in austenitic stainless steel

6.3 TYPICAL DEFECTS

6.3.1 Castings and ingots

Shrinkage during solidification and cooling in the mould is the main problem in the production of shaped castings and ingots for further working, and this can produce shrinkage cavities such as piping and hot tears.

Interdendritic shrinkage in shaped castings, such as valve bodies, produces small cavities that may interlink. If this occurs the casting is said to be 'spongy' and would weep on pressure testing. Sand inclusions or dross inclusions may be trapped in the casting, and the evolution of dissolved gases during solidification may give rise to a rounded form of porosity 'blow holes' if they are small, or 'pin holes'. Porosity resulting from shrinkage is readily detected by radiography.

6.3.2 Defects in wrought products

Many of the defects in wrought products, such as plate, sheet, tube, rod and wire that are manufactured from cast ingots; result from defects in the original ingot. Internal defects in an ingot, such as secondary pipes, usually do not oxidize during heating or hot-working, and will weld up to 'heal' completely when the ingot is rolled or otherwise hot-worked. Any external defect, such as primary pipe is open to the atmosphere and, will be oxidized during heating and will not heal during a working process and will give rise to defects, such as laminations, in plate or sheet (Fig. 6.6). This is probably the most common defect found in flat products, such as plate, sheet or strip, and is considered unacceptable, particularly if it emerges at edges that are to be welded. Fortunately, laminations can easily

Fig. 6.6 An 'old sixpence' split down the middle by severe lamination in the cupronickel strip from which it was made

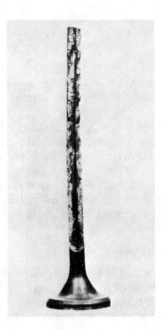

Fig. 6.7 Motor car engine valve that has split along the piping defect running down the centre of the bar

be detected by ultrasonic inspection. In bar, the primary pipe will elongate to become a 'pipe' down the centre of the bar (Fig. 6.7).

If the surface of an ingot is not dressed, that is cleaned and any defects such as tears removed, before working, surface discontinuities become rolled into the surface of a bar or tube and appear as a 'seam' or a 'lap'. In extrusion, the back end may contain a central defect rather like a pipe, but in this case the oxide has been gathered from the surface of an improperly dressed billet and forced into the centre of the extrusion by the metal flow during extrusion. Surface discontinuities and non-uniform deformations can produce laps in forging (Fig. 6.8).

An unusual defect is 'cupping', which may be produced during drawing when the material is insufficiently ductile to perform the deformation required of it as it passes through the drawing die (Fig. 6.2).

6.3.3 Heat treatment defects

Under certain conditions, the heating or cooling of both large and small pieces of metal can result in the formation of cracks (Fig. 6.9). Even a ductile metal such as aluminium can be made to crack by cooling it rapidly.

Cracking due to hydrogen and to transformation from one phase to

Fig. 6.8 Well oxidized forging lap in steel

Fig. 6.9 Quench crack in steel tools

another can also be produced by heat treatment.

Grinding cracks (Fig. 6.3) can be classed as heat-treatment cracks, in that they are produced as a result of localized rapid heating and cooling of the surface during grinding. They can easily occur if the supply of coolant is inadequate.

6.3.4 Welding defects

Welding still remains a difficult process to control, and in view of the fact that it involves rapid heating and cooling together with fusion and in some cases deformation of the metal, it is not surprising that many of the defects already discussed can occur. Such defects as heat-affected zone cracks, solidification cracks (hot tears), porosity and slag entrappment are commonly produced if proper welding procedures are not carefully instituted. Additionally, defects, such as lack of fusion and lack of penetration, maybe produced. These are both very serious defects and must be located by NDT in all critical structures.

6.3.5 Defects produced in service

Where components or structures are inspected periodically, defects that have developed as a result of the service conditions can be located by the use of NDT techniques. Without the use of NDT such defects only become apparent when a component fails by fracture. Three of the principle types of defect developed during service are fatigue, stress corrosion and creep cracks.

Fatigue cracks
   These are by far the most common defect to develop in service. The cracks are initiated at a free surface as a result of the action of repeatedly applied stress. Bolt holes, sharp changes of cross-section and similar aspects of the geometry of a component cause stress concentration and are typical sites for fatigue crack initiation. Fatigue cracks can be very 'tight' cracks and are not normally visible and can only be revealed by the use of a suitable NDT technique.

Stress corrosion cracks
   These result from the combined action of stress and corrosion (Fig. 6.5). A certain corrosive environment will not attack every material in this way, but most materials are susceptible to attack by one or more corrosive media.

Creep cracks
   Only lead and tin of the commonly used industrial metals and alloys creep at ambient temperature. For steels, the temperature must usually exceed 500°C before creep becomes significant. Accordingly it is normally only in plant such as steam and gas turbines, operating at high temperatures that such cracks are experienced and must be found. The cracks usually appear at the surface and can be located sometimes by good visual inspection, since the cracks are often wide, or by the use of a suitable NDT technique.

6.4 NDT TECHNIQUES

Over the last 50 years many NDT techniques have been developed to augment visual inspection, but only five of these have seen widespread use:

  (i) radiography — finds internal and surface flaws of a 3-D character
  (ii) ultrasonics — finds internal and surface flaws of a 2-D character
  (iii) magnetic particle inspection — finds surface and subsurface flaws
  (iv) eddy current techniques — finds surface and subsurface flaws
  (v) dye penetrants — only finds flaws at the surface

6.4.1 Radiography

This technique utilizes X-rays or γ rays to penetrate the solid components under inspection. Both rays are short wavelength electromagnetic radiation and their penetrating power and energy is inversely proportional to their wavelength. Short wavelength radiation with high penetrating power can be used to radiograph 10 ins (250 mm) of steel, and is known as 'hard' radiation. Long wavelength radiation of low penetrating power is said to be 'soft'. The wavelength of X-ray techniques can be controlled to match the component under inspection, but no such control is possible when a γ ray source is used. Hence X-rays are to be preferred for all high quality work, but γ rays have an important secondary role to play.

X-radiography
The heart of an X-ray unit is the Coolidge tube, which was first produced in 1913 and is shown diagramatically in Fig. 6.10. This consists of a heated filament in an evacuated glass envelope. The heated filament (cathode) emits electrons that are accelerated towards a tungsten target (anode) by a potential difference of usually more than 100 kV. There are two controls in the Coolidge tube:

  (i) filament current — this controls the filament temperature and hence the number of electrons emitted
  (ii) potential difference — the number of kilovolts between the filament

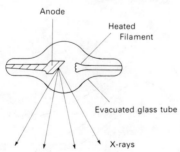

Fig. 6.10 Coolidge X-ray tube

and the tungsten target controls the speed with which the electron strikes the target

High-speed electrons produce harder X-rays than lower speed electrons. About 1% of the kinetic energy of the electrons is converted into X-rays. The rest appears as heat, and hence the target must be designed for rapid heat extraction. This is usually achieved by embedding the tungsten in a massive copper heat sink and pumping cooling gas around the exterior of the glass envelope that forms the tube.

The electrons striking the tungsten target knock out the orbiting electrons from the tungsten atoms. When these excited electrons drop back into their appropriate shells, X-rays are generated. The minimum wavelength of X-rays generated, i.e., the highest energy is given by the Duane–Hunt relationship:

$$\text{min wavelength } (\lambda_{\min}) = \frac{12.35}{V} \times 10^{-4} \text{mm} \tag{6.1}$$

where V is the applied voltage across the tube. The relationship is illustrated in Fig (6.11).

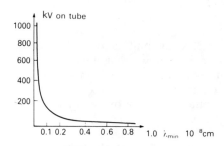

Fig. 6.11 Voltage on Coolidge tube against minimum wavelength of resultant X-rays

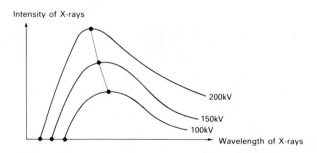

Fig. 6.12 Variation of intensity of X-rays with wavelength. (Note that the wavelength of maximum intensity shifts to lower wavelengths as voltage increases)

The intensity of the X-ray beam varies with wavelength as shown in Fig. 6.12. The range of wavelengths follows from the fact that the accelerated electrons have a range of speeds. Because the X-ray beam contains a range of wavelengths it is said to be 'white' by analogy with white light.

Unlike light, X-rays cannot be focused by lenses and ideally in order to obtain good resolution the X-rays should be emitted from a point source. Focusing the electrons onto a point on the tungsten target is possible, but would result in drilling a hole through the tungsten because of the intense heat generated. Accordingly, in a 220 kV tube, the electrons are focused onto a small area (which might be 2.5 mm × 2.5 mm) on the tungsten target.

γ-Radiography

Before 1948, radium was the main source of γ-rays used in industry, but this has changed with the advent of the nuclear reactor. Now the large neutron flux available in reactors has made available a wide range of radioisotopes. The nuclei of the atoms of these isotopes are unstable and disintegrate, emitting γ-rays. Their activity is measured in curies. (The new recommended SI unit is the becquerel (Bq); $1Bq = 2.703 \times 10^{-11}$ curies.) 1 curie produces an activity of 3.70×10^{10} disintegration per second, which is equivalent to the activity generated by 1g of radium.

Once a nucleus has disintegrated it cannot do so again, and therefore the activity of the isotope falls with time in the form of an exponential decay function. This leads to the concept of *half-life*. Half-life is the time taken to reduce the activity of a radio isotope to half its original value. Several isotopes are listed in Table 6.1 with their half-lives and the thickness of steel that they can readily radiograph.

Table 6.1

Isotope	Half-life	Thickness of steel (mm)
Cobalt-60	5.2 years	30–150
Iridium-192	75 days	15–65
Thulium-170	129 days	1–10
Tantalum-182	115 days	30–150
Caesium-137	8 years	20–70
Ytterbium-169	31 days	40–65 (approx)

Thus a range of thicknesses of steel can be inspected by having several different isotopes available.

The activity of the isotopes cannot, as with X-rays, be switched off and so they are housed in lead or depleted uranium containers when not in use. Thus an isotope can be taken out on site safely housed in its container and when required to produce a radiograph it can be removed from the container ready for use. The intensity of γ-ray sources is relatively small compared with typical X-ray installations and exposure times of several hours are required. The intensity can only be increased by increasing the size of the source, and this leads to loss of definition on the resulting radiograph. However for a given intensity, sources with shorter half lives, are smaller.

Cobalt-60 is the most commonly used isotope because its use enables relatively large thicknesses of steel to be penetrated that would otherwise require very expensive X-ray equipment to achieve the same result.

Absorption of radiation by solids

As X-rays and γ-rays pass through solids they are absorbed to some extent as depicted in Fig 6.13.

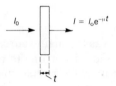

Fig. 6.13 Absorption of X-rays

As the thickness t increases, the radiation that comes through falls off exponentially. μ is the linear absorption coefficient, and depends on the material and the energy of the radiation.

The practice of industrial radiography

The set-up used to radiograph a weld is illustrated in Fig. 6.14. Where a cavity is present the radiation passes through a smaller thickness of material and hence arrives at the film as a more intense beam. When the film is subsequently developed the cavity will appear as a dark spot on the radiograph. The 'darkness' of the radiograph is known as its density. If light is passed through a radiograph, as in Fig. 6.15 the density can be measured by suitable instruments.

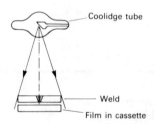

Fig. 6.14 Radiography of a weld

It is the density difference between one spot and another on the radiograph that enables defects to be seen. Obviously good viewing conditions, good eyesight and experience on the part of the inspector are prerequisites of good radiography.

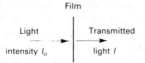

Fig. 6.15 Definition of film density

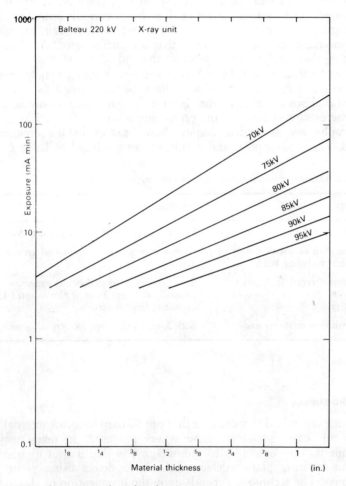

Fig. 6.16 Typical exposure chart for aluminium

A typical exposure chart is shown in Fig. 6.16, in which the exposure is the product of current (in mA) and the time of exposure (in min) i.e., mA. min. The chart is for a given film density and film development conditions. All radiographs must have the image of an Image Quality Indicator on them if the radiograph is to be accepted by the various standards. The most commonly used IQI (or penetrameter) consists of a series of wires of increasing diameter made from a similar material to that being radiographed (Fig. 6.17). Most standards require that a wire which is no thicker than 2% of the thickness of the component being radiographed, is visible. This requirement does vary with thickness of the component under inspection, since it is easier to 'see' a wire of 2% of the thickness of a component 100 mm thick than for one 10 mm thick. The image quality indicator does not indicate the defect sensitivity, but it does serve to show that proper radiographic practice has been carried out. The skilled radiographer has many specialized techniques that can be used to overcome the numerous difficulties that are encountered. A useful book describing these techniques is listed at the end of this chapter.

It must be stressed that both X-rays and γ-rays are extremely damaging to human tissue, and some deaths have been recorded as a result of accidental exposure to radiation. Both techniques should only be carried out where safety can be of prime importance. The advantages of radiography are such that despite these hazards its use continues to expand. These advantages and disadvantages are listed in Table 6.2.

Table 6.2 Advantages and disadvantages of radiography

Advantage	Disadvantage
Can find internal and surface flaws	Planar defects are frequently not detected
Can be applied to a wide variety of components including welds	Cannot inspect thicknesses of steel greater than 300 mm
A permanent record of defects is produced	Rising cost of silver makes film expensive, especially th rge areas of film required for inspecting large structures
Interpretation is relatively easy	Both X-rays and γ-rays are very dangerous

6.4.2 Ultrasonics

This is the second NDT technique that can be used to locate internal flaws, but in this case, harmless sound waves are used. In many ways the technique is complementary to radiography, in that it is at its best when used for locating planar defects difficult to detect using γ- or X-ray techniques. The technique depends upon the propagation of elastic waves (sound waves) through solids and their reflection at defects. Ultrasound is usually defined as being high-frequency sound with frequences higher than 20kHz, which is considered to be at the top end of the audible range. In practice frequencies of 0.5MHz to 20MHz are used for flaw detection.

Transducers
These are the probes used in the detection of defects and utilize

Wire diameters

Wire no.	Diameter	Wire no.	Diameter	Wire no.	Diameter
	mm		mm		mm
1	0.032	8	0.160	15	0.80
2	0.040	9	0.200	16	1.00
3	0.050	10	0.250	17	1.25
4	0.063	11	0.320	18	1.60
5	0.080	12	0.400	19	2.00
6	0.100	13	0.500	20	2.50
7	0.125	14	0.630	21	3.20

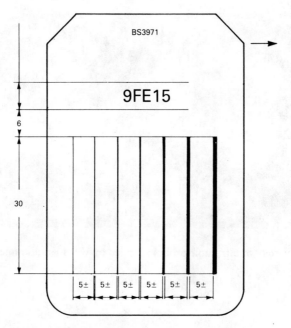

All dimensions are in millimetres.

Fig. 6.17 IQI from BS 3971 — wires 9–15 illustrated

piezoelectric or electrostrictive crystals that can be made to vibrate at the high frequencies required by applying an alternating electric field across them. Common materials used are quartz and barium titanate. These transducers convert the applied electrical energy into mechanical energy (sound), and vice-versa. A wide variety of probes is now available to tackle the many problems that can arise in inspection. Some such probes are illustrated in Fig. 6.18. If the crystal transducers are embedded in a block of perspex such that the sound wave enters a block of steel at an angle (Fig. 6.19), the sound beam is refracted as it enters the steel. If the angle of refraction is 60°, the probe is said to be a 60° probe for steel.

Fig. 6.18 A variety of ultrasonic probes supplied by Wells-Krautkramer

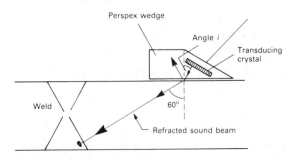

Fig. 6.19 Angle probe locating defect in weld

Five different types of sound wave can be used for detecting defects:

(i) longitudinal or compression waves — where tension and compression occurs in the same direction as the direction of propagation of the waves
(ii) transverse or shear waves — where the tension and compression occurs at right angles to the direction of propagation
(iii) surface waves (Rayleigh) — these ripple along the surface
(iv) plate waves (Lamb waves) — in thin plates the whole plate is made to vibrate
(v) bar waves — similar to (iv) for thin bars

The velocity of longitudinal waves in various materials is given in Table 6.3.

Table 6.3 Velocities of longitudinal waves

Medium	Velocity	
Air	3.3×10^2	ms^{-1}
Water	1.43×10^3	ms^{-1}
Steel	5.8×10^3	ms^{-1}
Copper	4.6×10^3	ms^{-1}
Aluminium	6.2×10^3	ms^{-1}

The velocity of shear waves in steel is 3.2×10^3 ms^{-1}, which is just greater than half the longitudinal wave velocity.

In the situation depicted in Fig. 6.19, the longitudinal wave travelling through the perspex is totally reflected at the interface with the steel, but the transverse wave, being of lower velocity, enters the steel at 60° to the normal in accordance with Snell's Law:

$$\frac{\sin i}{\sin 60°} = \frac{\text{Longitudinal velocity in perspex}}{\text{Transverse velocity in steel}} \qquad (6.2)$$

Angle probes are commonly used for weld inspection because of their ability to interrogate the whole of a butt weld by moving between full and half skip positions, as shown in Fig. 6.20

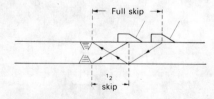

Fig. 6.20 Weld inspection using angle probes

The Ultrasonic technique

A typical ultrasonic defect (or flow) detector is shown in Fig. 6.21, and it will be seen that a cathode ray oscilloscope forms the main feature of the detector. The oscilloscope operates a time base on the x-axis, whereas the y-axis is activated when voltage is applied to the probe to produce the ultrasonic pulse or when voltage is generated in the probe when it receives a reflection of that pulse. Accordingly, the presentation of a small defect in a plate is as shown in Fig. 6.22. This indicates what happens in the most frequently used 'A' scan presentation using the pulse–echo technique. In this technique, small pulses of elastic energy are repeatedly sent out by the probe, and these are reflected back from defects and from the back wall of the plate. Because the pulses are of short duration, the same probe can be used to both transmit the signals and receive them. However, if the defects are near to the surface the defect echo becomes confused with the transmission pulse, and in this case separate transmitting and receiving crystals are necessary. By suitable calibration, the distance x_1 can be obtained by measuring x on the screen (Fig. 6.22). In this way the position of the defect is pinpointed. The height of the defect echo on the screen is a measure of the sound pressure received by the probe. The bigger the reflecting area of the defect, the bigger the sound pressure received by the probe and the bigger the defect echo on the screen of the ultrasonic set. However, estimating defect sizes from the height of the echo on the screen is very dangerous since large defects oriented at inappropriate angles can give small echos.

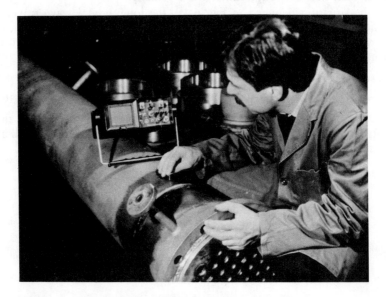

Fig. 6.21 Typical ultrasonic flaw detector

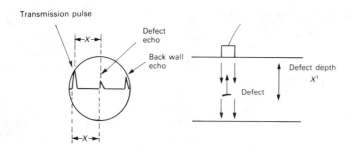

Fig. 6.22 'A' scan representation of a flaw in a plate

The interpretation of the signals that appear on the screen of the ultrasonic defect detector using 'A' scan is not easy, and inspectors need thorough training. Further, the inspection procedure can be slow and tedious. Some organizations have developed a system whereby probe movement is synchronized with the x/y axes of the screen. Brightness corresponds to echo strength. The inspector is then presented with a plan view of plate being inspected. This speeds up the inspection and possibly improves the reliability of the technique, but it does not gain more information than is already available to the skilled 'A' scan operator.

Examples of applications of ultrasonics

Ultrasound will travel large distances through good quality wrought metals and alloys and will readily pass through 300 mm of steel. The technique was originally developed for inspecting hair line cracks in steel forgings and it is still much used for this. In recent years with the increased use of welding as a means of fabrication ultrasonic inspection has expanded rapidly in this area. The welded structures used in the North Sea oil programme are now thoroughly inspected using ultrasonics, and cracks have been found in welds that radiography had deemed to be crack free.

The advantages and disadvantages of using ultrasonics are summarized in Table 6.4.

Table 6.4 Advantages and disadvantages of using ultrasonics

Advantage	Disadvantage
Finds planar defects readily	Defect sizing is presently still difficult
Portable equipment can be operated on site without difficulty	Needs skilled operators who have to be trusted
Can be built into automatic systems	No permanent record easily available
No danger from sound waves	

6.4.3 Magnetic particle inspection

This method is very commonly used in industry, but it is restricted to ferromagnetic materials, such as iron, nickel, cobalt and their alloys. Fortunately, most engineering steels, with the exception of the austenitic steels, are ferromagnetic.

Principles of magnetic particle inspection
   The technique depends on the fact that magnetic lines of force will leak out at a surface crack in a magnetized steel component as in Fig. 6.23.

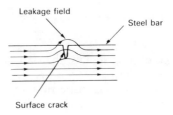

Fig. 6.23 Principles of magnetic crack detection

   This leakage field is attractive to any magnetic particles and these collect around the crack. When a suitably magnetized component containing a surface crack is covered with magnetic ink (a suspension of iron powder in paraffin) the iron particles collect at the crack. If the surface was bright with machining, the crack will show as a dark line (Fig. 6.24). For use with darker surfaces, the iron powder may be coated with fluorescent dye and the component has then to be viewed in a dark room using ultraviolet light. Cracks containing oxide and inclusions can be found, as well as

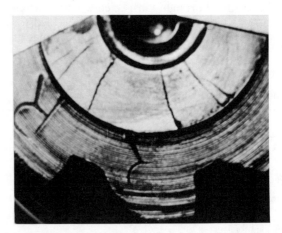

Fig. 6.24 Radial grinding cracks as revealed by magnetic particle inspection

defects just below the surface which generate sufficient leakage field. The technique is particularly useful in detecting small grinding cracks in ground components.

Equipment used in magnetic particle inspection

A simple way to magnetize a component is to use a magnet, and this method is sometimes used in the field. However, better results can be obtained by using Universal machines (Fig. 6.25). Such machines are capable of passing a heavy electric current through a component, thereby producing a magnetic field at right angles to the direction of current flow. Additionally, the machines contain powerful electromagnets that can produce a field at right angles to the one generated by the electric current. Defects are not detected if they lie along the direction of magnetization and so the use of Universal machines is justified. The magnetic particles can be applied after the field has been switched off — this utilizes permanent magnetism. Alternatively, the particles can be applied when the field is switched on, and this is the more sensitive technique.

Fig. 6.25 Crankshaft being tested by magnetic particle testing unit

Direct current of several hundred amps is expensive to produce so most machines use half-wave rectified a.c. However deeper defects are detected using d.c., whereas simple a.c. best detects defects confined to the surface. After inspection, demagnetization is usually required, particularly if there is to be no interference with magnetic compasses.

Applications of the technique

The technique is used mostly for routine inspection of machined steel components. Small cracks, such as grinding cracks, are easily found, as are inclusion stringers. However the technique also finds use in the inspection of large steel castings and forgings. The advantages and disadvantages of using magnetic particle inspection are summarized in Table 6.5.

Table 6.5 Advantages and disadvantages of magnetic particle inspection

Advantages	Disadvantages
Good workshop technique that does not require highly skilled operators	Relies on the integrity and eyesight of the inspector
Can be carried out rapidly	Can only be used on ferromagnetic materials
Finds inclusions as well as cracks	Does not find deep internal defects
Can find defects just below the surface	Gives no accurate indication of the depth surface defects
Can be recorded photographically	

6.4.4 Eddy current crack detection

If a coil energized with a.c. is placed close to a conducting metal, eddy currents will be generated within the metal. If the currents are strong, then the metal rapidly heats up by 'induction heating'. However, small currents can be equally useful to industry by providing a means of surface inspection, since the eddy currents will be smaller where defects are present at the surface. The magnitude of the eddy currents can be measured by a suitable detector coil. The situation shown in Fig. 6.26 could be used to inspect a bar for surface defects. The pick-up coil will detect any change in the field strength of the exciting coil due to discontinuities in the bar surface. The depth of penetration of eddy currents is usually quite small, and can be said to be inversely proportional to the square roots of frequency, magnetic permeability and electrical conductivity. Thus, low frequencies have to be used if deep penetration is required. Additionally the high magnetic permeability of steels makes it necessary to saturate steel bars magnetically before eddy current inspection. This can readily be done by using magnetizing coils. In order to get good defect resolution, frequencies of the order of 10 000 Hz are used.

Not all the coils are of the encircling type shown in Fig. 6.26. Hand-held probes can also be used, and after suitable calibration these may give the

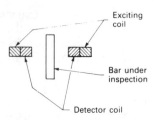

Fig. 6.26 Eddy current inspection

depth of a surface crack, providing it is not much deeper than 2 mmm.

The technique lends itself to automation, and can be used to inspect ball joints for surface cracks as well as tubes and bars. Thin-walled tubes can be fully inspected by encircling coils, but thick-walled tubes will require an additional inspection of the bore. This can be achieved by inserting suitable coils into the bore. Of course, piping defects at the centre of bars cannot be found by eddy currents, but this defect can readily be found by ultrasonics. The advantages and disadvantages of eddy current techniques are summarized in Table 6.6.

Table 6.6 Advantages and disadvantages of eddy current techniques

Advantage	Disadvantage
Lends itself to rapid automatic inspection (bars can be inspected at speeds in excess of 30 mph)	Does not readily find surface pits and pores
With suitable calibration, crack depths can be found	Only finds defects at or near the surface
Works very well on regularly shaped components, such as bars and tubes	Inspection of welds or irregularly shaped forgings or castings is tedious
Can be made to work on all electrically conductive materials	Signal/noise ratio is poor if surface is rough

6.4.5 Dye penetrant techniques

These techniques require very little skill to carry out or interpret, and the equipment involved is inexpensive. Consequently, they are extensively used for the detection of surface flaws in components and structures. Essentially the techniques require the penetrant to penetrate into the surface cracks and then to ooze out of the cracks at a later stage, thereby indicating the presence of the cracks.

The stages in using a water-washable penetrant are:

(i) apply the penetrant after degreasing and allow time for penetration
(ii) rinse the penetrant from the surface with water
(iii) apply a developing powder, usually French chalk, consisting of fine spherical particles. This encourages the dye to seep out of the crack spreading through the interstices of the developing powder
(iv) visual inspection is the final stage. If the dye is of the fluorescent type then this is done in a dark room using ultraviolet light. If the penetrant is one that requires emulsifying, then the emulsifier is added after penetration and before rinsing. Certain other penetrants are removed by solvents instead of by water rinsing

The fluorescent penetrants that require emulsifying are claimed to be the most sensitive, but much depends on the cleanliness of the operation and the penetrating time. Penetration times up to 1 hour are used to locate fine defects. The technique is very valuable in locating fatigue cracks since, except in highly exceptional circumstances, these always start at the surface. The advantages and disadvantages of the technique are summarized in Table 6.7.

Table 6.7 Advantages and disadvantages of the dye penetrant technique

Advantage	Disadvantage
Inexpensive and requires only a low level of skills for operation	Surface defects filled with dirt or oxide may not be found
Gives very clear indications of fatigue cracks	Only finds surface defects
Can easily be used in the field	Gives no reliable indication of crack depths

6.5 MINIMUM DETECTABLE DEFECT SIZE

It has been possible for several years to estimate the critical defect size, a_c, for defects in various structures. The critical defect size is that which if exceeded will eventually lead to failure of the component or structure. For example, the critical defect size for cracks in the main welded seams of pressure vessels may be 100 mm in the 'as-welded' condition and 120 mm in the stress-relieved condition. However the presence of hydrogen in the heat-affected zone of a weld may reduce the critical defect size to only 1.25 mm (Whittaker, 1972). Where fatigue stressing can extend small crack-like defects the initial critical defect size is often less than 1.0 mm. Table 6.8 shows the allowable slag and porosity in a steel butt weld under fatigue conditions.

Table 6.8 Slag inclusions and porosity levels allowable in steel butt welds under fatigue conditions (after Harrison, 1969)

Maximum design fatigue stress (MNm^{-2})	Maximum length of slag inclusions (mm)		Uniform porosity that can be tolerated (% vol)
	As-welded	Stress relieved	
0–140	No max.	No max.	20
140–220	No max.	No max.	20
220–280	10	No max.	8
280–370	1.5	6	3

Stress corrosion is another mechanism that can extend small defects so that they become critical. Table 6.9 indicates that a nickel–chromium–

molybdenum steel (SAE 4340) can have a critical defect size as low as 0.1 mm when stressed at 275 MNm^{-2} in salt water.

Table 6.9 Critical defect sizes in stress corrosion cracking (after Whittaker, 1969)

| Metal
18% Nickel Maraging steel | K_{ISCC}
(MNm$^{-\frac{3}{2}}$) | Applied stress
(MNm^{-2}) | Critical flaw size
in salt water (mm) |
|---|---|---|---|
| 18% Maraging steel | 40 | 275 | 5 |
| SAE 4340 | 14 | 275 | 0.1 |

These examples indicate that in general cracks and crack-like defects are far more dangerous than slag and porosity and that in certain circumstances even quite small defects can be potentially dangerous.

It is now necessary to discuss whether modern NDT techniques are able to locate the types of flaw that fracture mechanics has shown to be important.

6.5.1 Radiography

It is now well known that the depth of a defect in the through thickness direction is much more important than the length of the defect in determining whether or not it is dangerous. Unfortunately the depth of a defect is very difficult to measure by radiography and there is no image quality indicator that measures the crack sensitivity of radiography. According to the theory developed by Pollitt (1962), however, the crack sensitivity and wire sensitivity can be related for low-energy radiation as follows:

$$pw = 1.2d^2 / (1 + \frac{d}{U_{\mathrm{T}}})$$

$$(6.3)$$

where p is the depth of unit crack length, w is the width of unit crack length, d is the diameter of wire — just visible on the radiograph (see Fig. 6.17) — and U_{T} is the total unsharpness on the radiograph.

Suppose a 1% wire sensitivity is achieved in the radiography of a steel butt weld in plate 25 mm thick. Then the diameter (d) of the smallest wire whose image is just visible is

$$d = \frac{1}{100} \times 25 \text{ mm} = 0.25 \text{ mm}$$

Now using equation (6.3) and assuming U_{T} to be 0.25 mm:

$$pw = 3.75 \times 10^{-2} \text{ mm}^2$$

so if $w = 0.01$ mm, then $p =$ depth $= 3.75$ mm. Thus a defect 10^{-2} mm wide must be more than 3.75 mm deep if it is to be just visible under these conditions. Hanstock (1971) gives another estimate of the minimum detectable cross-sectional area:

$$pw \simeq 4 \times 10^{-5}t^2 \qquad\qquad (6.4)$$

where t is the thickness of the material being radiographed. If $t = 25$ mm, then $pw \simeq 4 \times 10^{-5} \times 25 \times 25$. If the width $w = 10^{-2}$ mm then:

$$p = 4 \times 625 \times 10^{-3} = 2.5 \text{ mm}$$

This value of the minimum crack depth detectable is slightly more optimistic than that obtained above according to Pollitt's equation.

Holt (1962) has determined experimentally the minimum crack sizes that are detectable, and his results are shown in Fig. 6.27.

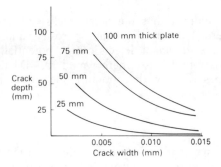

Fig. 6.27 Minimum detectable crack depths using radiography

If a plate 25 mm thick has a crack width of 10^{-2} mm Fig. 6.27 predicts a minimum detectable crack depth of approximately 3 mm, which is in good agreement with the two calculated values above. These data are for the optimum situation where the X-rays run down the crack, i.e., the X-ray beam is parallel to the plane of the crack. If the beam makes an angle with the plane of the crack, then it is much more difficult to detect the cracks. Referring again to the work of Pollitt, it is possible to calculate the angle at which a crack disappears. This is compared with experimental data in Table 6.10.

Table 6.10 Angle of disappearance of small cracks in radiography

Voltage	Steel thickness	Crack dimensions		Angle of disappearance	
(kV)	(mm)	W (width) (mm)	P (depth) mm	Observed	Calculated
200	25	0.6×10^{-2}	2.5	6°	$5\frac{1}{2}$°
330	37	0.025	3.1	5°	5°
330	37	0.025	4.7	15°	$12\frac{1}{2}$°
400	50	0.025	3.0	5°	4°
400	50	0.025	4.5	10°	$6\frac{1}{2}$°

In welds, cracks may often be orientated at greater angles to the X-ray beam than those listed; especially if there are lamellar tears. Radiography is therefore not very suitable for fine-crack detection, but it is the best technique for detecting small rounded defects. Recent work by Hudgell and Tickle (1979) has shown that radiography can detect gas pores as small as 0.1 mm diameter in the butt-fillet welds of the type used in the manufacture of steam generators for the prototype fast reactor at Dounreay.

6.5.2 Ultrasonics

There is no doubt that the ultrasonic technique is far better at detecting cracks and crack-like defects than radiography. The author has shown by laboratory tests using surface wave probes on polished surfaces that it is possible to detect fatigue cracks that are so small that they cannot be seen by the naked eye. In practice, however, the situation is not so encouraging as indicated by the recent PISC tests (Nicholls, 1980). PISC is the abbreviation for the Plate Inspection Steering Committee of the Nuclear Energy Agency's Committee on safety of Nuclear Installations. The work involves 35 teams from 10 different countries. The plates which were inspected were made in USA from a Mn/Mo weldable steel SA-302 Grade B and were 200 and 250 mm thick. Two of the plates were butt welded by the electro-slag or submerged arc processes and the third one contained a nozzle which was welded in by the manual metal arc process. The samples contained implanted and natural defects plus some drilled calibration holes.

The inspection was carried out by teams of skilled and certified operators under very nearly ideal conditions. There was no clad overlay on the plates and the surface condition was reasonable. A conventional 'A' scan technique was used using compression wave probes and 45° and 60° angle probes. The results of this exercise have proved disappointing. Single vertical crack-like defects 30 mm deep were only correctly reported by about half the teams, but 60 mm deep defects were correctly reported by most teams. Multiple defects were found difficult to assess by all the teams, even when these constituted a significant flaw. Defects near to the surface proved difficult to locate. When the teams were allowed to use their own particular inspection techniques instead of the ASME XI Code required initially, the situation improved. The reasons for the disappointing results are many, but amongst them may be listed:

 (i)   poor standardization of probes and electronics
 (ii) reflection from defects is very dependent on angle of attack
 (iii) poor surface condition of the plate giving rise to poor coupling
 (iv) inadequate operator training and integrity

Examining large areas is obviously tedious so that it is important to determine the critical areas using fracture mechanics. When the critical defect size turns out to be small ultrasonic inspection needs to be most thorough. Even so it is unlikely that there will be a high probability of

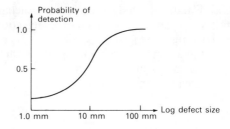

Fig. 6.28 Marshal Committee's assessment of the probability of detecting small defects

detecting small defect as shown by Fig. 6.28 which indicates the Marshall Committee's assessment of the situation (Nicholls, 1980).

Figure 6.28 clearly indicates that it is only possible to find defects bigger than about 40 mm with a high degree of certainty. It must also be remembered that inspection of welded fabrications should be carried out before and *after* stress relief since residual compressive stresses can make certain defects transparent to ultrasound (Ibrahim and Whittaker, 1980). In short, present technology cannot guarantee to find the smaller defects with certainty, but if adequate precautions are taken, all large defects should be detectable.

6.5.3 Sensitivity levels for other methods of inspection

Magnetic particle and dye penetrant techniques are dependent on the ability of the inspector to see the defects with the naked eye. It is usually claimed that defects down to 0.25 mm in length can be detected reliably by the magnetic method, although they might need to be 6 mm long before being reliably detected by dye penetrants. Much depends on the nature of the defect and the nature of the surface being inspected. Neither magnetic crack detection nor the dye penetrant technique can measure the depth of cracks that emerge at the surface. However if it is assumed that the depth is equal to the length of a crack, this will usually err on the safe side probably by a factor of two.

Eddy current techniques can measure the depth of surface breaking cracks if it is assumed that the crack plane is at right angles to the surface. Fatigue cracks in rivet holes can readily be measured over a depth range of 0.125–1.75 mm in aluminium alloys. The skin effect limits the useful depth of penetration of eddy currents, but with luck, ultrasonics might take over where the cracks are too long for eddy currents to be suitable.

Table 6.11 summarizes the situation for both surface and internal defects.

6.6 NDT AS AN AID TO PROCESS CONTROL

Many plate mills, bar mills and tube mills now incorporate NDT systems

Table 6.11 Defect sizes that can be detected by various techniques

Technique		
Surface cracks	Minimum length that can readily be detected and measured (mm)	Minimum depth that can readily be detected and measured (mm)
Ultrasonics	5	2
X-rays	12	Cannot measure depth
Eddy current	0.8	0.125–1.75
Magnetic particles	0.25	Cannot measure depth
Penetrants	8	Cannot measure depth
Internal cracks	Smallest diameter (mm) readily detectable	Size estimates
Ultrasonics	2.5	Various methods available
Radiography	Not detectable unless beam virtually parallel to crack	Not possible

into their quality control organizations. It is useful to consider how NDT was used in the modernization of one of Britain's steel plants, and this is well described in a paper by Boyes and Eldridge (1976). The plant modernization involved installation of new melting casting and rolling processes. The main products of the works were tube and engineering rounds, and it was estimated that the direct costs of failing to meet customer requirements averaged 5% of sales value. It is generally accepted that this direct cost can be doubled by other expenses incurred by the company and its customers. These losses occur despite expenditure of a further 2.5% sales value on quality appraisal and failure prevention. The figures suggested that savings could be achieved by better control of quality, and it was considered that automatic NDT could prove a valuable tool in this direction. It soon became clear that regular monitoring and feedback of two features of heavy round sections were required:

(i) internal quality assessment for both micro and macro defects
(ii) surface quality assessment to measure the depth of defects

Products requiring critical inspection

It was decided that rounds 75–165 mm in diameter provided the bulk of the high quality orders, and accordingly the system was designed to inspect these for alumina aggregates (down to 400 μm in size) and to discriminate between inclusions, porosity and cracks. It was also necessary to meet surface standards requiring that there should be no surface defects of depth greater than 0.9 mm.

Inspection techniques used

A conventional ultrasonic testing scheme to test bars for internal quality was estimated at £1 000 000 (1976) values) and proved too costly. The main expense was in bar handling associated with the need to allow bars to cool in order to carry out ultrasonic inspection. Accordingly a non-contact ultrasonic technique (EMA) was developed. This made warm ultrasonic testing at the mill finishing end feasible. Pilot equipment was built to operate in-line to inspect 165 mm diameter bar for internal defects at a much reduced cost of £20 000 (1976 values).

Attention was then turned to automation of the surface inspection techniques and an eddy current system was developed. This operated off-line at room temperature and was installed close to the EMA tester. Another EMA head was used in this off-line position together with sophisticated signal processing equipment to give more detailed information on internal quality. The in-line non-contact ultrasonic test facility has been able to work at speeds up to 55 m min^{-1} and at temperatures up to 500°C. Defects, such as pipe, cracks and inclusions, can be located under these operating conditions, and the results obtained from EMA have been confirmed by normal manual ultrasonic testing.

The off-line test installation is standardized against an axial 1.6 mm artificial flaw and eight surface artificial flaws within a range of 0.40–1.25 mm in depth. The performance of the off-line EMA and eddy current systems was checked in the laboratory on bars containing both internal and surface flaws, such as seams, cracks and laps, and found to be satisfactory. In the early days of the installation up to 200 bars/week could be tested off-line. An example of what might be achieved in automatic ultrasonic inspection systems is shown in the defect discrimination 'tree' of Fig. 6.29 taken from the paper by Boyes and Eldridge (1976).

The above is an example of what can be done in automatic inspection. Ultrasonic, eddy current and magnetic inspection techniques lend themselves to automation, but radiographic and the other techniques are difficult to automate though in the case of radiography some progress has been made (Lee-Frampton, 1980).

Magnetic methods using flux leakage techniques can be automated to inspect surface defects in bar and test speeds up to 90 m min^{-1} can be achieved on bars from 20 mm to 450 mm diameter. Defects 0.6 mm deep and 0.1 mm wide can be detected in hot rolled-billets with excellent signal/noise ratios.

It will be seen that automated NDT techniques are largely confined to the inspection of regular shaped components, such as bars, tubes and plates, but for these shapes the automated techniques are well developed and are an essential part of modern production plant.

A new technique of detecting surface defects in bars has recently been described (Luz, 1980).

This technique is suitable for testing steel bars with as rolled surfaces and uses a strong a.c. (1–30 kHz) magnetizing field, plus suitable test probes to detect the flux changes associated with surface flaws. The so-called Rotoflux instrument is capable of inspecting black or sand-blasted

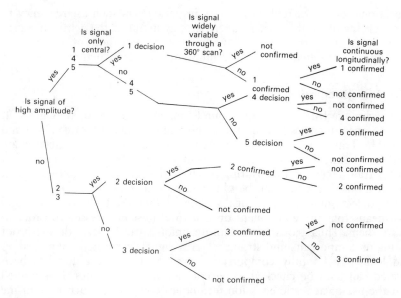

Defect code

1 Pipe	4 Single cracks
2 Porosity	5 Multiple cracks
3 Inclusion aggregates	

Fig. 6.29 Defect discriminating 'tree'

bars of 25–120 mm diameter in-line. Test speeds of 0.3–2 ms^{-1} can be achieved to give a yearly throughput with 3 shift operations of up to 300 000 tons. Defect discrimination is claimed to be better than when using magnetic inks at defect depths of 0.25 mm.

6.7 THE CONCEPT OF CONTINUOUS SURVEILLANCE

In many components and structures, such as pressure vessels, oil rigs and aircraft, there is a need to monitor the performance continuously to make sure all is well. At present this is not readily possible, and a vessel's integrity might have to be assured by NDT during regular shut-down periods. If the same function could be achieved by some kind of continuous monitoring system then the costly shut-down periods might be avoided. Three possible means of providing such continuous monitoring are:

  (i) vibration techniques
  (ii) ferrography
  (iii) acoustic emission (stress wave emission).

Vibration techniques rely on changes in parameters such as frequency to give warning of trouble, whereas ferrography monitors the debris in engine oil to predict wear problems. Only acoustic emission of these three techniques, can detect defects and pinpoint them in structures.

6.7.1 Acoustic emission

This technique listens in to the 'sounds' made by defects as they grow. It does so at frequencies beyond the audible range usually between 100 and 300 kHz. Transducers similar to those used in ultrasonic flaw detection are used to monitor the sounds emitted from a defect. By using an array of at least 3 transducers and measuring precisely the time of arrival of a signal, the location of a defect can be obtained by triangulation techniques.

The defects must be acoustically active if they are to be detected, and this means that the component or structure must be stressed in order to locate them. Thus acoustic emission techniques can be used during proof testing of components and structures. Any signals generated are picked up and located by a fully computerized system. The defect that has been located can then be checked using conventional ultrasonics. It is argued that the use of acoustic emission techniques could save much time in the inspection of complex structures and components, but snags can arise when no emissions are detected. In this case, the question arises: does it mean that no flaws are present or is there an instrumental failure? Also, emissions from more than one defect confuse the picture and it is possible for defects that remain 'quiet' during one mode of stressing to become active in a different stress situation.

These problems have not, however, deterred acoustic emission enthusiasts, and several schemes for continuous surveillance by acoustic emission have been suggested and are now used. These rely on the concept that should flaws become active or develop during service, the sounds they produce can be monitored and, for example, a pressure vessel could be shut down before it fails.

Another example of the use of acoustic emission is in weld monitoring immediately after welding. Cracks that develop in welds after welding can produce strong acoustic signals, and these can be picked up as acoustic emissions. Again the problem is to be sure that the weld is up to standard if no emissions are picked up.

There is no doubt that acoustic emission, which is still a relatively new technique, is here to stay, though its impact on the defect detection scene will not be as great as it was originally expected.

6.8 SUMMARY

This chapter has set out to describe the flaw types that can be found in engineering components and structures, and how to locate them non-destructively. Five techniques are in regular use throughout the world, and these are being continually improved. The use of fracture mechanics can indicate the size of flaws that must be located and a large part of current work concerned with NDT, particularly in ultrasonics, is being devoted to the sizing of defects. In parallel with this are the many developments in automatic inspection systems and these have been given a boost by the advent of the microprocessor.

In this one chapter it has not been possible to describe all aspects of NDT, but it is hoped that the more important areas for the engineer have been covered. It finally should be stressed that modern inspection is an interdesciplinary affair and the NDT engineer, materials engineer and design engineer must work closely together in the early stages of any project. Otherwise components and structures may be designed that cannot be inspected and designing for easy and valid inspection should now be a prime requirement for design engineers.

SUGGESTIONS FOR FURTHER READING

J.W. Boyes and P.E. Eldridge (1976). The role of NDT in the modernization of a steelworks. *Iron Steel Int. Dec.*

A. Butler (1979) *Metall. Mat. Technol.* 11, 29–34.

R. Halmshaw (1982) *Industrial Radiology*, Applied Science Publishers, London.

R.F. Hanstock (1971) *Pressure Vessel Engineering Technology* (ed. R.W. Nichols), Elsevier, London, p. 372.

J.D. Harrison (1969) *Welding Inst. Res. Bull.* 9, 227.

D. Holt (1962) *Limitations in Detecting Crack-like Defects by Radiography*, British Engine, Boiler and Electrical Insurance Co. Ltd., Technical Report, Vol. 4, pp. 71–87.

R.J. Hudgell and J.H. Tickle (1979) *Developments in Pressure Vessel Technology*, Vol. 2, Applied Science Publishers, London, p. 77.

S.I.Ibrahim and V.N. Whitaker (1980) *Brit. J. NDT* 22, 286–290.

J.B. Lee-Frampton (1980) *Brit. J. NDT* 22, 228.

H. Luz (1980) *Brit. J. NDT* 22, 232–235.

R.W. Nicholls (1980) *Metal Construction* 12 (5), 226–229.

L.R. Parkes (1979) *Metall. Mat. Technol.* 11, 13.

C.G. Pollitt (1962) *Brit. J. NDT* 4, 71.

V.N. Whittaker (1972) *NDT* (now *NDT International*) 5, 92–100.

CHAPTER 7

Economics

7.I INTRODUCTION

The *Deacon's Masterpiece* emphasizes the technical objective of design and materials engineers, and the earlier chapters have emphasized the increasing complexity of the problems facing those engineers in designing components today. So far it has been primarily the technical aspects of material properties that have been dealt with, but underlying all that has been said has been the requirement that what is being designed shall, when produced, be saleable.

Figure 0.1 indicates the cyclic nature of design and infers the need for a successful compromise to be reached between all the many factors involved. Not indicated directly in Figure 0.1 is the fact that technical parameters and fabrication costs may have to be weighed against more imponderable problems, such as safety, hygiene, availability of materials or equipment, legal and sometimes political and strategic factors.

7.2 CAUSES AND ORIGINS OF SELECTION PROCEDURE

It will be appreciated that the underlying reason for undertaking the design evaluation procedure is for one or several of those outlined in Table 7.1.

Table. 7.1 Causes and origins of selection procedure

Creation of a new product/function
Fuel cells, solar energy cells
Skate boards
Nuclear fuel can
New materials for aesthetic effects
North Sea oil rigs
Gas turbine engines
Novel electric batteries

Modifications to meet new or anticipated service conditions
Reactors, storage and other vessels and pipelines for chemical processes
Aluminium beer cans
Desalination equipment
Aircraft and gas turbine developments
Industrial gas turbine engines
New marks of automobile engines

Cost reduction
Nodular cast iron to replace forged steel
Weldments to replace castings
Various metal plastic interchanges for many ancillary motor car fittings
Pipeline distribution services for water, gas, effluent, etc.
Use of adhesives for metal joinings

Materials giving inadequate service
Corrosion problems arising with condenser tubes, motor car silencers, car bodies and
   chemical plant
Lightweight armour plate
Surgical implants

Uses for new alloys/materials and new treatments to improve properties
Highly formable superplastic alloys
Memory metals
Composite materials
Ceramics
Plasma and ion bombardment treatment of metals

Certain of these imponderable factors can only be considered in the context of the locality in which the item being dealt with is to be manufactured or used. Certain of them can only be considered in the context whether they are meant for a military or civil application. But others must be considered in all cases.

Two such factors that are of prime importance and that have almost an overriding effect on the economics of production and therefore on saleability are the cost of the material from which the component is to be made and the cost of the energy used in its production.

These two factors, material and energy cost, are discussed further in this chapter, along with brief comments on other factors that may contribute to, or influence, the cost of production.

7.3 METAL COSTS

7.3.1 Base metal prices

The first contributing factor to the cost of a component is the cost of the raw or virgin metals from which it is made. Virgin metal prices are determined fundamentally throughout the world by supply and demand. In the case of iron in the form of pig iron, the wide range of alloy steels and cast irons, the price is fixed by the intense competition between the producers, and a similar condition applies to other metals such as aluminium, cobalt and tungsten.

Certain other metals such as copper, tin, lead, zinc and nickel, are bought and sold daily by dealers on the London Metal Exchange, and the transactions made at the daily market fix the price for those metals.

Typical prices for metals are quoted in the daily press and in journals such as *The Metal Bulletin* or *American Metal Market*.

7.3.2 Conversion costs

It is seldom, however, that an engineer can use these specific prices directly, for the bulk of metal is not used in the pure or virgin state, but is normally alloyed and then worked into the multitude of useful shapes useable by the designer and known as semi-finished products.

Thus most suppliers of these semi-finished products will quote single prices that are composed of two parts: firstly a part based on metal mixture values; and, secondly, a part which is the added value or conversion cost incurred in alloying and working the metals, and this also includes a yield factor. This yield factor is the ratio (%) of the acceptable material produced to the start material. The metal mixture value is normally derived directly from the values of the component metals in the alloy, and may or may not include a valuation depending on the proportion of scrap used during manufacture. Examples of the multiplying factor used to cover the added value or conversion costs for various brasses are given in Table 7.2. Should a substantial change occur in the metal prices quoted on the London Metal Exchange, then if it is upward the multiply factor would decline, and conversely, if downwards, it would increase. It is these prices and values that an engineer must concentrate on and determine when comparing one material with another.

Table 7.2 Examples of multiplying factor in pricing wrought brass products

Alloy	Form	Basic price	Multiplying factor	£ tonne^{-1}
58/39/3 Cu Zn Pb rod 1″ dia. × 10′	Extruded	Mixture value est. 653.5	× 1.54	1006.0
58/39/3 Cu Zn Pb section 2″ × ⅜″ × 10′	Extruded	Mixture value est. 653.5	× 2.01	1313.5
63/37 Cu Zn wire 0.064″ dia.	Wire in coil	Mixture value est. 682	× 2.07	1412.0
63/37 Cu Zn 4′ × 2′ × 0.080″	Sheet	Mixture value Est. 682	× 2.25	1534.5
63/37 Cu Zn 6″ × 0.028″	Strip in coil	Mixture value est. 682	× 2.20	1500

Metal prices at 11 Aug. 1980 (£ tonne^{-1}):
  Copper 895
  Zinc   318
  Lead   335

Companies that use large tonnages of any of these products can frequently arrange purchases at favourable terms through long-term contract. Nowadays, even these contracts are priced subject to the daily London Metal Exchange prices, but proportionate discounts may be obtained subject to the size and rate of delivery. It is also possible to minimize the risk of price fluctuations in metal by a hedging operation on the London Metal Exchange.

Of the alloying component of the conversion costs, even quite small amounts of other elements, whether specifically added or controlled, e.g., titanium or niobium added to stainless steels, or oxygen content controlled in copper or carbon content reduced in extra low carbon stainless steels, will each augment the price significantly.

7.3.3 The effect of the production of scrap

Another aspect of the final true cost of buying metal into any engineering works is the net difference in value between the cost of the metal bought in and the price obtained for selling the metal as scrap arising in the course of manufacture. If the identity of the scrap is maintained by careful segregation it can command a good price relative to that paid for it originally.

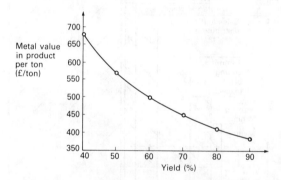

Fig. 7.1 Relationship between yield and metal value in product

The significance of the effect on the cost of the metal in a component of the percentage of scrap produced during the production of the component is well illustrated in Table 7.3 and Fig. 7.1. The Table shows the effect of yield, that is the percentage of the metal originally purchased actually used in the final product, on the cost of the metal in that product. The example taken assumes a purchase price for the incoming metal of £360 tonne[-1] and a selling price of the scrap produced in the form of, say, trimmings and turnings of £150 tonne[-1].

It will be seen that as the yield increases from 30% to 90% the cost of the metal actually used in the component drops from £850 tonne[-1] to £384 tonne[-1], that is it more than halves. Figure 7.1 illustrates this factor graphically.

7.3.4 Significance of overall yield on cost

In the process of selecting the metal or alloy best suited to the design and

Table 7.3 Yield sensitivity of product metal value

Yield %	Redeemed value of scrap (£)	Inherent metal value of product (£)	Total metal value in product (£)	Metal value of product per tonne (£ tonne^{-1})
30	10 500	10 800	25 500	850
40	9 000	14 400	27 000	675
50	7 500	18 000	28 500	570
60	6 000	21 600	30 000	500
70	4 500	25 200	31 500	450
80	3 000	28 800	33 000	412
90	1 500	32 400	34 500	384

Assumptions
Ingoing forging stock at £360 tonne^{-1}
All scrap after forging at £150 tonne^{-1}

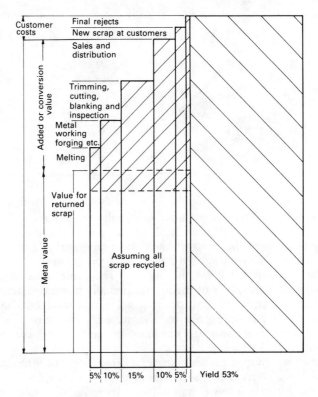

Fig. 7.2 Build-up of total cost from metal values, added values, yields and recycling scrap

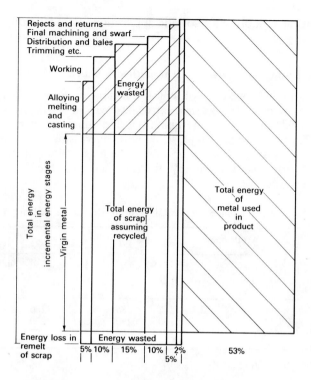

Fig. 7.3 Build-up of total energy used in the manufacture of a metal product

final manufacturing requirements it is equally important to consider the shape and form of the metal as purchased. Clearly this has a very significant effect on the final total yield of metal used in the finished part, for the nearer the as-purchased metal is to the final form the less scrap is produced. In addition, the nearer the original material is to the final form the less is the energy required to manufacture the component.

A schematic diagram of the significance of yield in cost and energy terms at each step in manufacture of a wrought metal is shown in Figs. 7.2 and 7.3. The decrease in yield or scrap losses are arbitrary values, but can be considered fairly average in the manufacture of semi-finished products. The scrap losses in manufacturing a component from a semi-finished product can vary from 95% in the case of an aluminium wing spar for an aircraft, chemmilled from solid rectangular plate, down to perhaps 1% for cold-headed and roll-threaded bolts.

With metal in the cast or extruded form, the significant steps involving scrap losses or rejections are less for castings or extrusions, which are inevitably closer to the final form required but, nevertheless, the same principles hold.

The forementioned general principles hold true whatever the metal or material. There are, however, other aspects of producing metal in a finished form suitable for a specific application, and in return these affect the costs.

7.3.5 The pricing structure for semi-finished products

Discounts on the basic price of semi-finished metal products in the form of plate, sheet, strip, rods, sections, extrusions, wire and tube can be obtained as mentioned earlier for large quantities and running orders. Tables 7.4, 7.5 and 7.6 illustrate this and also show the increasing cost as the product is reduced in thickness, width or cross-section as a result of the additional work put into the product by the supplier to achieve these smaller dimensions.

Table 7.4 Basic prices (£ tonne^{-1} at 2 April 1978) for steel cold-rolled strip

| | Width (mm) | | | | | | | |
| | 15–20 | | 25–50 | | 50–75 | | >300 | |
Thickness (mm)	H	S	H	S	H	S	H	S
4–3.5	300	325	261	285	256	280	242	265
1–0.8	302	329	268	290	247	270	245	268
0.26–0.24	358	444	356	409	336	383	354	397

H = hard   S = soft

Extras and allowances (£ tonne^{-1})
Rimmed Steel 6.70–12.00 according to C and Mn contents
Killed steel where applicable 12.5–18.80 according to C and Mn contents
Copper, nitrogen and niobium extras for 2.0 to 8.90
Deep drawing quality 15.00
Grain size controlled 4.20
Quantity > 20 tonnes   − 1.50
        20–5 tonnes Basis
        5–3 tonnes + 5.00
        2–1 tonne +15.00
Non-standard coils from Basis to +45.00 according to dimensions of coil
Cut to lengths 13.00 to 55.00 according to dimensions
Fine tolerances 10–25 £ tonne^{-1} according to thickness, width, edgewise bow, length and flatness
Finish   matt   4.00
        plating   20.00
Testing to BS 1449 Pt. 1 '72.   Basis   AQD (W) Defence Standard Issue   20.00

If the quality of either the surface or the interior of the product is second grade or inferior, further discounts can usually be obtained.

However it may be that the component to be produced demands a higher quality or that better sales and better service can be obtained by using materials having high quality finishes and for these extra prices would be charged. These increases would cover such aspects as high quality polished or abraded surface finish, special surface treatments,

Table 7.5 Aluminium extrusions

wt/ft	Classes			Cost (£ tonne⁻¹)		
	A*	B*	C*	A	B	C
0.05	1.75	2.75		3828	3828	3960
0.50	5.00	7.5		2508	2574	2740
5.00	15.00	24.00		2442	2508	2574
50.00	50.00	79.00		2442	250	2574

Classes are where perimeter of extrusion section is number of inches.
C is > limits of B
Quantity extras and allowances (£ tonne⁻¹)

> 10 tonne	−15	Extras
1–2 tonne	Basic	A.I.D. inspection 20–40 £ tonne⁻¹
0.5–1	+10	Controlled stretching 20 £ tonne⁻¹
0.25–0.5	+15	
0.05–0.15	+150	Die changes 100–400 £ tonne⁻¹

Approximate comparisons assuming aluminium is about £600 tonne⁻¹. Price at 15 Oct. 1980 is £750 tonne⁻¹

Table 7.6 Basic price (£ tonne⁻¹) of aluminium coil and flat-rolled product (sheet and strip coiled and flat)

	Thickness (in)					
	Specs. 1A to N53			Specs. N54 to HS30		
	0.1	0.048	0.012	0.1	0.048	0.012
Coil width (in.)						
0.5– 1.0		2376	2706	3102	3168	4884
3.0– 6.0	2180	2240	2508	2970	3036	4554
36.0–48.0	2180	2240		2904	3036	
Flat sheet width (in)						
>1.5		2904	3498	3432	3630	
3.0– 6.0		2442	3036	3036	3234	
36.0–48.0		2310	2706	2904	3036	

Quantity extras and allowances

>20t	- 3%	>0.25t	1250 £ tonne⁻¹
> 1t	- 2%	0.05t	1910 £ tonne⁻¹
> 1t	Basic		

Approximate comparisons assuming aluminium is about £600 tonne⁻¹. Price at 15 Oct. 1980 is £750 tonne⁻¹

exceeding the normal limits of size or requiring smaller dimensional tolerances. Even minor points, such as special flatness or minimum bowing of strip, will call for extra charges.

There will be some variations in the proportion of such charges between the different metals. Because aluminium and its alloys have a relatively low yield point and high elongation on stretching, it is almost universal practice, with no additional cost to the customer, to stretch the finished product, whether extruded or flat rolled, to improve the straightness and flatness. This is not so with steel and copper rods and sections, which are straightened either by reeling or drawing or sometimes both, and this will result in extra charges.

The importance of considering these aspects at the engineering design stage will become clearer as the various aspects are discussed further.

7.3.6 Quantity

Typical discounts on quantity are usually quoted in the commercial literature, but can often be improved by negotiation between purchasing officer and salesman, particularly if the quantities are sufficiently large to result in a regular and steady delivery rate so that a contract including delivery rates can be agreed. Unless quantities are very large, one must bear in mind the stranglehold that a single supplier might have on production, should supplies fall short or stop for some unforeseen reason. In this context, there has been a vast expansion in the supply and selling of metal via metal stockholders or warehouses strategically located in appropriate cities and towns. Some of these warehouses and supply stores or depots are tied to a particular manufacturer, but probably more than half are independent and thus in a position to obtain metal supplies from sources throughout the world. They are, of course, inclined to charge somewhat more than the direct supplier, but will usually supply in smaller lots. They have the advantage of being able to buffer very significantly the disruptive effects of trade disputes that may interrupt supplies obtained directly from a manufacturer.

7.3.7 Quality

The main criterion with regard to quality is that supplies are obtained from a reputable supplier where quality assurance procedures are faultless. There is, however, the possibility that significant discounts in price can be obtained for off-quality products, which yet will, in certain applications, give perfectly satisfactory manufacturing and service performance. Such off-quality material may have any one, or combination of, minor faults. It is often more expensive and certainly involves more energy to consign the second-grade material to the melting pot than to sell it for use where it will behave perfectly satisfactorily.

There is a well-developed trend for quality assurance procedures to be made mandatory, partly because (i) safety factors in design are being reduced, (ii) the need for increased reliability in service, and (iii) greater stringencies and demand for increased safety. Quality assurance is discussed in Chapter 5. It must be realized that such quality assurance procedures must be paid for and may result in a surcharge on the product. Charges of £20–£40 tonne^{-1} are cited in Tables 7.3, 7.4 and 7.5 for steel and aluminium products.

Working metals at elevated temperatures normally involves considerably less energy than working at room temperature. Furthermore, the properties in the hot-worked condition may be only slightly different from those in the cold-rolled and annealed condition and metals 'as hot-rolled' are lower in cost than when 'as cold-rolled'. However, steels and copper base alloys in the hot-rolled condition do retain mill scale on the surface and in the case of aluminium a smooth mill finish, and an additional cost may be required to improve or paint the surface.

Another aspect of quality of particular significance in steels is the presence of impurities as tramp elements, or as inclusions. For many applications, the presence of these impurities will not affect the performance of a component, but it must be remembered that, as described earlier, under fatigue or high or low temperature conditions of service, steels in particular can fail under stresses below design stresses. Typical price increases due to control of copper, nitrogen and niobium are given in Table 7.4.

7.3.8 Surface finish

Normally extra charges are made for any metal surfaces that are improved or cleaned using mechanical, ultrasonic, vapour cleaning or chemical methods. Mechanical cleaning can be abrasive using an abrasive belt or wheel, or air blasting with an abrasive media such as sand, and may be used to give a clean matt finish or, in some cases, to remove surface contaminants, such as oil or furnace-deposited films such as the carbon film formed inside copper tubes.

A clean metal surface might be desirable for corrosion resistance, as in stainless steels, cupronickels, nickel silvers, or for aesthetic appearance, or as a base for adherence of paint or plastic surface films. Clean surfaces are also essential for many joining processes, such as soldering, brazing or welding.

There is also a wide range of electrochemical treatments. These include electrolytic cleaning and polishing, the deposition of protective films electrochemically, such as chromium or nickel plating, or the anodizing of aluminium and magnesium alloys. Such treatments may greatly increase the corrosion resistance or, as with anodized aluminium finish or chromium plating, increase the hardness and wear resistance.

The range of surface treatments and finishes available is now so great that authorities in this field must be consulted to ensure that the most appropriate treatment is selected.

7.3.9 Limits and tolerances

As the design engineer will know, these terms all relate to the dimensions and true geometric form or shape of the product, and are defined in practice as the smallest or largest dimensions in three directions that are required to be met to ensure that a component produced using them will perform satisfactorily and will assemble correctly with other components. If very fine tolerances are demanded, the cost of meeting these increases as is indicated in Tables 7.4, 7.5 and 7.6.

7.3.10 Flatness, edgeways bow and thickness

The stresses and strains involved in the working or deformation of wrought metals are inherently greater than any stresses or strains to be involved in service. The mechanical forces needed to deform metal, particularly in thin gauge, are so high that distortion of, say, the rolls in a rolling mill may occur and it is in fact almost miraculous that flat, straight strip, sheet or bar is ever produced. The working operations are so severe that they can be regarded as a valid acceptability and reproducibility test or form of quality assurance.

Probably the most difficult dimension to satisfy is that of thickness across the width of a metal strip. The producer of the strip will intentionally produce strip with a slight 'middle', that is, the strip will be thicker at its centre than at its edges. This middling is required to ensure straightness for coiling and slitting, and engenders good shape or flatness when the strip is uncoiled.

The amount of 'middle' must be minimal to achieve the producer's objectives for, if too great, distortion and bowing of the strip will occur if it is slit into smaller widths.

This variation in thickness across the width of the strip is paralleled by similar variations in the dimensions of sheet or sections and, if tight tolerances on these variations are required by the designer, clearly additional costs will be incurred.

These examples of additional costs that may be incurred, if special requirements are placed on the supplier of a metal, could be extended to include examples where strict control of grain size, grain orientation, presence of burrs on cut edges, hardness of the metal and other factors were required, and all such special requirements will increase the cost to the user. Such cost increases must be justified economically.

7.4 ECONOMICS AND THE PRODUCTION ROUTE

In the early days of the use of metals they were formed to the final shape required by casting to that shape, making bar and cold- or hot-forging to shape, or making bar and cutting away metal by a machining process until the required shape was obtained.

Today there are many more ways of producing the required shape, and new methods or variations of old methods are continually being developed.

The first stage is usually to produce the required metal in the molten form, and it is in the subsequent stages that processing and economics of the processing will vary.

7.4.1 The casting route

Clearly the closer to the final shape that results when the molten metal is solidified, the lower are likely to be the processing costs. Much attention has been paid therefore to achieving this objective, and many casting-to-shape processes have been developed, amongst which are sand, shell, gravity or pressure diecasting, and investment casting. Each of these variants has advantages and limitations; thus, diecasting is normally limited to alloys of comparatively low melting point, such as those of aluminium, zinc and lead. Investment casting is normally used with more exotic alloys based on nickel, chromium and cobalt, and so on, where the melting point is high but the higher cost of the process can be tolerated commercially.

However, for metallurgical reasons, explained in earlier chapters, cast metals do not show the property values or the consistency of properties that are available with wrought materials but, by close attention to the maximum property values required and a knowledge of the potential weaknesses of cast metals, they can be exploited in many engineering applications.

7.4.2 The wrought metal route

If a cast component is not acceptable the design and materials engineers must explore alternative methods of production, assessing the economics of each.

The starting material, with the exception of the powder metallurgy processing of electrorefining routes, to be mentioned later, is likely to be bar, rod, sheet, strip, tube or wire produced by the hot- and cold-working of ingots cast from the molten metal.

The alternative methods of production include:

machining
forging
stamping, hot or cold
cold or hot deep drawing
welding
pressing

or a combination of these.

The designers must examine each potentional route, its effects on the properties and performance of the product, and the economics involved.

7.4.3 Other production routes

As an alternative to taking the molten metal and forming it to the desired shape using one of the casting techniques, the metal can be converted to a powder form and this powder pressed to the required shape in a die and sintered (that is heated to a temperature at which the metal particles weld together). The formed metal may be used in that condition or given a final cold pressing or coining operation, which further solidifies the powder compact and corrects any dimensional irregularities.

A further method of forming metal to a required shape is to

electrodeposit it on to a mandrel or mould of the required shape. This procedure is only adopted to meet special requirements, such as in waveguides, where accurately maintained internal dimensions are required.

The component being considered by the designers may in itself decide the production route to be followed; thus certain alloys, which are the only materials suitable for use in a specific application, can only be used in a casting process and certain other alloys cannot be cast to shape. The designers must therefore, throughout the design exercise, not only consider the strength and dimensional requirements to be met, but also the production route and the economics both of the material selected and that production route.

7.5 SCRAP METAL

Whether one is designing a machine or structure in metal irrespective of the expected life, there remains an inherent value in terms of metal cost and total energy, which should not be ignored.

In considering this matter, it is important to differentiate between two different kinds of scrap, namely, new scrap and old scrap. Although there may not be too much difference compositionally between them, for any particular alloy, nevertheless, it must be made quite clear that new scrap is that which arises solely in customers' works as scrap from the process of manufacture and therefore should retain its alloy identity and be free from contamination. Old scrap, however, is scrap arising from metal that has been in service, then discarded. It's identity, composition and freedom from contaminants is much more doubtful. It therefore usually commands a lower metal value, unless it has been melted, analysed and adjusted to a known specification.

If the life proves to be short, say two to five years, then it is important to ensure that the metal used is easily recyclable with the minimum extra energy and with the minimum of loss in metal value. This latter can occur either by contamination by impurities, or by metal losses in demounting the component, or refining or remelting. However, if the life-time of the metal is long, such as with a component used in a power station, a bridge, a ship or a civil aeroplane, then at the end of its life the value of the metal will almost certainly be enhanced as a result of inflation. Again, a significant contribution to the cost of a metal component is the cost of the energy used in its manufacture, and this is a valuable asset since the total energy content was incurred when the energy costs were considerably lower than when the metal is scrapped.

Those metals and materials that are not durable or cannot be recycled will slowly prove themselves uneconomic because both the material value and total energy expended will be lost forever. Such inevitable economic forces will call for reassessment, in detail, of many wasteful uses of metals and materials, some principles of which are illustrated diagramatically in Figs. 7.4 and 7.5.

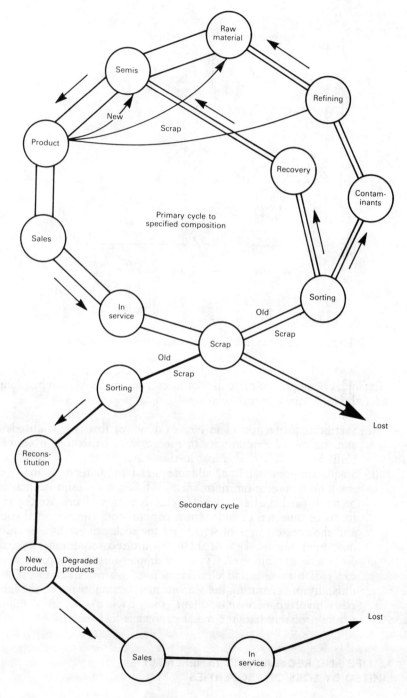

Fig. 7.4 Recycling of metals or materials (schematic diagram)

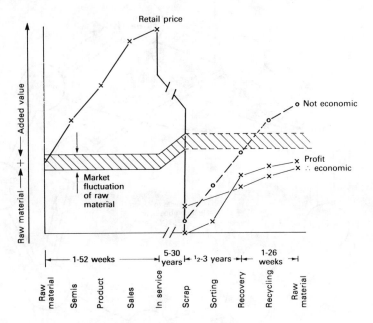

Fig. 7.5 Hypothetical life-cycle and values of a metal or material

Examples of single life-cycle uses of metals and materials can be grouped under the following general heatings:

(i) Sacrificial protection. The best example of this is galvanized iron and the use of zinc anodes to protect steel structures in seawater. Half the world's zinc is lost in this way.

(ii) Small components. If a valuable metal or material is made into small units to economize on metal value, it may result in such units being discarded after one life-cycle. Examples of this are the many forms of tungsten carbide which contain both tungsten and cobalt, and the greatest uses of which are for rock drilling as tips and for machining metals. Few of these tips are recovered and recycled

(iii) Non-demountable uses. Typical examples in this group are motor-car radiators or small electric motors, where either the physical difficulty of separating the various metal components or the labour costs involved necessitate them going back to the metal refinery, where probably half the metal content is lost.

7.6 LIFE AND RECYLABILITY INHERENTLY LIMITED BY LOSS OF PROPERTIES

Fortunately, metals do not suffer from this debility, except by fracture mechanisms or corrosion. Even so, they can be remelted and the original properties restored. This is not so with many non-metallic materials, such as wood, rubbers and several plastics, in common use. On the whole, they are single life-cycle materials and if they are worth collecting, can be recycled, but, because certain mechanical and physical properties may be lost, their value may be reduced and they can only be used as secondary materials having poorer properties.

Usually however, costs are so heavily against the single-life materials that they are destroyed by burning or dumping. Considerable material value and total energy is therefore lost to mankind in a single life-cycle usage. This observation cuts at the very roots of the so-called 'waste economy' or 'throwaway society'.

7.7 ECONOMICS AND THE FUTURE

This chapter has been headed, 'Economics', but since it also deals with future trends, an attempt has been made to consider some new concepts in design philosophy that will have to be increasingly recognized as additional determinant factors in optimizing design during the next 30–40 years.

The fundamental reason for introducing these new factors is that sources of energy are diminishing and energy will increasingly prove to be the dominant cost component of most finished products if the total energy required for their production is considered. Unfortunately, this total energy is not likely to be truly reflected in the costs of the product until it becomes really dominant and unless and until energy supplies are not subject to subsidies, restraints and restrictions by various countries where such materials are processed.

Engineers in general and design engineers in particular will increasingly, therefore, have to consider the resource limitations both of energy and of some materials and metals in determining the design and manufacture of the overall product, or assembly. There is no doubt that these longer-term concepts are at variance with most present-day commercial and industrial attitudes and customs. These are mainly the throwaway society, limited life, and fashion changes.

Nevertheless, the energy and resource limitations outlined later make it seem inevitable that the world will increasingly be driven to using additional and new design criteria.

There are also other more immediate aspects that will probably tend to dominate the engineers' short-term thinking, and design and manufacture. These are the ramifications of product reliability laws, whether national or international, both for the individual and for the organization. In addition, avoiding polluting the environment can substantially increase the manufacturing costs of certain metals or alloys and in some cases, where throughput is not sufficient to carry the costs of antipollution measures, will result in cessation of manufacture, as with zinc smelting in California.

Most engineers are required to reduce costs in everything they design or make. However, apart from simple manufacturing costs, designers will be increasingly obliged to consider other important criteria, which may be one or other of the following:

  (i)   weight reduction
 (ii)   reliability of all components parts and whether for a limited or 'indefinite' life, e.g. bridges
(iii)   interchangeability
 (iv)   fail/safe
  (v)   aesthetic values
 (vi)   recyclability
(vii)   welding (joinability)
(viii) assembly
 (ix)   environment (temperature, atmosphere, liquid)

Simplicity is the basis of good design and in most engineering machines and structures costs are increased more than proportionately by the use of a large number of sub-parts. A complex sub-assembly requires many assembly procedures and, in turn, high labour costs.

7.8 ENERGY AND COSTS

It is certain that market forces will in the future predetermine the usage of energy and materials, but this will only happen when the energy costs become a major proportion of the cost of any end-product. In some cases, energy is already 30% of the value but, since the energy expenditure on incoming raw materials used earlier in the processes are not quantified in separate financial terms, most managers and directors believe that the energy contents in their part of the process only represent 10% to 15% of the total cost. Rigorous energy auditing would reveal a much higher value, and indeed one author suggests that energy is one fundamental economic measure. One could argue that energy and labour are almost the total cost of any metal or material.

Publication of total energy-utilization data in metallurgical and material processing operations has been widespread in recent years. However, such data are in rather an early stage of compilation. In some cases there are discrepancies of ×2 to ×3 between values in different countries and between different firms. Recent work in the UK iron foundries and aluminium industry has revealed that energy usage in routine metallurgical establishments, measured over extended periods, is up to double the hitherto assumed values. Difficulties have been experienced in assessing true works production data in total energy increments at each step, partly because metering of individual production units on a complex works site is not carried out, and partly because yields of individual products and the degree of occupacity of the plant influence the values. Other errors stem from various assumptions as to the overall energy uses on a site. The overall estimate of the total energy that has been already incurred by an ore as raw material or concentrate or virgin metal when it has arrived in the UK is also not readily calculated.

Considerably more detailed work and agreed conventions will be necessary on an international basis before total energy data on materials is truly comparable. The energy audit series of investigations by government departments are probably the most accurate detailed reviews of energy usage in the UK, but so far they have only covered a limited range of industries.

In the interim, that is the next 20 years, it seems that total energy content in a product will not be accurately reflected in its cost. This is partly because the total energy content of a raw material is not known or appreciated, and partly because various energy sources are so vastly different in basic cost and the true cost hidden by state subsidies, and partly because energy auditing as yet cannot fairly apportion the energy used on a works site to each product group, nor assess how energy use varies with output.

There is therefore considerable justification of the need to assess accurately and discuss the relevance of the total energy content of the major tonnage materials of the world.

Two of the aspects of total energy auditing that need general agreement and implementation are related to scrap recycling and the justification of manufacturing one life-cycle only and/or dissipated materials.

The value of the energy content of material produced by recycling scrap consists of two components:

(i) the inherent total energy originally used to make it the original material
(ii) the 5% to 10% additional energy required to recycle the scrap ready for re-use

Most firms would like to assess it at (ii) only, thereby gaining (i) for nothing. On the other hand, most scrap metals have an intrinsic financial worth that is fairly close to their raw-metal values with the exception, perhaps, of scrap iron and steel, whose worth is artifically low.

This financial value of scrap metal contains the value of the total energy content of its original extraction and manufacture. It would be anomalous to ignore this total energy content when the scrap is recycled, since lower total energy values would thereby be obtained for the new processed product.

The contrary view that on second and subsequent reprocessing much less additional or process energy is used per unit of product is attractive for short-term marketing, but cannot be tenable on strict energy-auditing standards.

The second problem is the vexed question of composites. These are defined for this instance as any geometric arrangement of two or more materials that cannot easily be separated into their component materials without excessive expenditure of energy or with low yields. This wide definition would cover such products as steel-cored tyres, glass-reinforced plastics, bimetals, motor-car radiators, honeycomb structures, laminates and carbon filaments. In all such cases, the main reason for manufacture is that in service these composites contribute significantly to savings in operating energy or longer life.

Since by their very nature and manufacture, composites are unlikely to be recycled, the energy used in making them must be justified. This can only be done by equating the energy used in their manufacture against the operating energy likely to be saved over the life-time of the product. At the very lowest these two values should be equated, that is, total energy used in complete manufacture to finished product = process operating or recurring energy saved over the average life of the product in service.

If the operating energy saved is greater than this, then the use of the composite is justified. If it is less, then the composite product must be justified for other valid reasons, such as:

  (i) life of component is increased
 (ii) the energy to make it is less than if a single recyclable material is used. This is unlikely
(iii) a unique low-energy technology is involved with no alternatives, e.g. microcircuitry and solid-state materials
 (iv) maintenance is significantly reduced

7.8.1 Total energy per unit of property

Although the total energy consumed per kilogramme of finished material can be evaluated, such information does not convey the inherent value of the product to prospective users. The concept of assessing the value of the product in total-energy terms for the range of properties being considered is of equal if not greater relevance. An outline of the properties of tensile strength, modulus of rigidity and fatigue strength for some common metals and materials is given in Table 7.7, together with their specific energy, i.e., total energy per kilogramme of material, from which date the energy per unit of property can be readily calculated. As would be anticipated, total energy criteria throw a completely different light on the true values of some materials to mankind. For example, timber uses far less energy for a given tensile strength (26–55 kWh MN^{-1} unit of strength), than any other material. Reinforced concrete is attractive at 145–250, followed by steels at 125–350, whereas cast irons can vary from 300–1825 in the UK, according to the energy efficiency of the manufacturer! The newer materials — aluminium, plastics and titanium — use total-energy contents for each MN unit of strength of 400 for duralumin, 700 for other aluminium alloys, 700 for titanium, and 500–2000 for plastics, depending on the type of polymer and whether energy content of the feedstock is included. Roughly the same order of merit obtains for modulus of rigidity and fatigue strength.

In Chapter 1, Table 1.1 considered the same materials and their costs. By a similar set of calculations, one can evaluate the cost per unit of property. For many of the materials this results in a somewhat similar

Table 7.7 Energy consumption relation to material properties

Material	Tensile strength (MN m⁻²)	Modulus of rigidity (MN m⁻²)	Fatigue strength, (MN m⁻²)	Density (kg m⁻³)	Specific energy (kWh kg⁻¹)	Total energy per unit of MN m⁻¹		
						Tensile strength	Modulus of rigidity	Fatigue strength
Cast iron								
(Castings)	400	45 000	105	7 300	4.0–16.0	73–292	0.65–2.60	278–1112
Steels								
En1 low-alloy free cutting bar	360	77 000	193	7 850	16.0	349	1.63	651
EN24 1.5Ni–1Cr–0.25Mo bar	1000	77 000	495	7 830	16.0	125	1.63	253
Stainless 304 18Cu–8Ni sheet	510	86 000	250	7 900	32.0	229	2.94	487
Non-ferrous metals								
Brass 60Cu–4OZn bar	400	37 300	140	8 360	27.0	565	6.05	1612
Aluminium alloy sheet	300	26 000	90	2 700	79.0–83.0	711–747	8.2–8.6	2370–2490
Duralumin sheet	500	26 000	180	2 700	79.0–83.0	427–449	8.2–8.6	1185–1245
Magnesium alloy bar	190	17 500	95	1 700	115.0	1029	11.17	2058
Titanium 6A1–4V bar	960	45 000	310	4 420	155.0	715	15.2	1520
Plastics								
Propathene GWM22	35	1 500	7.5	906	20.0–40.0	517–1034	12–24	2400–4800
Polythene L.D. XRM21	13	84	3.25	920	15.0–30.0	1062–2124	165.0–330	4246–8492
Rigidex 2000	30	1 380	4	950	15.0–30.0	475–950	10.3–21	3563–7126
Nylon 66 A100	86	2 850	20	1 140	50	663	20	3400
PVC (R)	50	1 680	12.5	1 400	20.0–50.0	560–840	17.0–25.0	2240–3360
Reinforced concrete								
Beam	38	10 000	23	2 400	2.3–4.0	145–253	0.5–0.96	240–417
Timber								
Hardwood	14	4 500	6	720	0.5	26	0.08	60
Softwood	5	2 000	3	550	0.5	55	0.14	92

ranking as when they are ranked in order of energy for a given tensile strength, that is, reinforced concrete, steels and cast irons, followed by a range of polymers with aluminium alloys competing with plastics. However, it will be seen that there are a few drastic shifts in the ranking Table, for example, titanium.

This type of analysis can be extended to cover other properties of engineering and life performance. Further, by a system of weighting and scaling the significance of a property in the overall performance sense, it is possible to determine the cheapest material in total energy or in cost terms for any predetermined combination of properties.

7.9 AVAILABILITY — ENERGY

An important factor to be considered is that as a resource becomes more difficult to discover and win, so the total energy required per tonne of finished product rises. This is well exemplified in the case of lead (see Fig. 7.6), where it is indicated that below a certain lead content in an ore body, the energy cost for its recovery becomes prohibitive and consequently lead

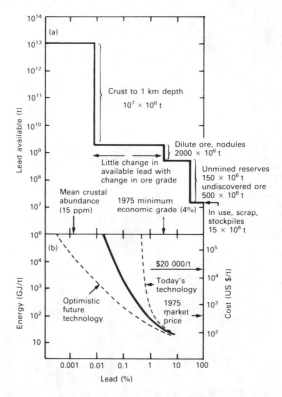

Fig. 7.6 Availability cost of lead v. ore grade

supplies corresponding decrease. Countering this difficulty, it is practical to recycle lead already in service, and this can be done with the expenditure of $\frac{1}{10}$–$\frac{1}{20}$ of the energy required to produce the lead originally.

Although most estimates in strictly reserves terms show a levelling off in the world's energy reserves nevertheless, as prices rise, energy demands tend to fall and other resources come into use.

In any effort to conserve energy, materials and, therefore, costs, the following major factors must be emphasized:

(i) The total energy data is only just becoming available and collated and its significance appreciated. Despite the difficulties in assembling data and vagaries of that data, it is fundamental and clearly of increasing cost significance

(ii) Ultimate yield of the finished product can vitally affect total energy and cost. Scrap wastes energy and material

(iii) The higher the value of a specific property the more efficient it usually is in energy terms but costs often increase more than proportionately

(iv) Costing and energy estimates must be for as near finished shapes as possible to include the bulk of energy and labour in the yield and complete added value

(v) To augment further the use of S.G. irons, apart from exploiting the low total energy values outlined earlier, it seems that determined efforts should be made to recycle the lower grades of steel and iron scrap in the cast iron cycle, comparable with the practices in the copper and aluminium fields.

7.10 AVAILABILITY — METALS/PLASTICS

Unfortunately, probably eleven of the metals that are now in use will decline in availability and are likely to become prohibitively expensive. These will probably be antimony, cadmium, cobalt, copper, lead, mercury, platinum, silver, tungsten, tin and zinc. What is not so readily agreed and is certainly more difficult to estimate is the time span of the so-called 'half-life' before resources are exhausted, but it could be another 30–50 years for certain of them. Fortunately, the remaining metals and non-metallic materials are much more abundant as can be seen in Fig. 1.5, but all plastics are at present derived from oil. It requires 2–5 tonnes of oil for every 1 tonne of plastic; therefore prices are more directly linked to oil prices than any other material.

7.11 STRATEGIC METALS

The winnable reserves of the world are distributed haphazardly and this can, from time to time, make some metals apparently of strategic importance. It is then that the recycling of old scrap metal and the use of strategic stockpiles of metal become important.

7.12 LONGER LIFE AND BETTER RECYCLING IS ESSENTIAL

One immediate and relatively easy way to conserve energy is to aim to double the life of all products and then double it again. There is no doubt that the growth in the use of many metals and materials will slowly level out, owing to a variety of circumstances. The reasons are that the richer grades of ore are being exhausted and it becomes increasingly more costly, in both energy and labour terms, to mine and process the lower grade of ore available. The same problem will also confront plastics, for oil is both the energy source and the feedstock for those materials.

Other materials that are much more readily available will supplant those that become scarce in many applications, a continuation of the competition for usage that has been intensifying over the past 200 years. In the present situation of no growth in availability for many materials, the availability and ease of recycling old scrap becomes vital, and it is essential for designers to bear in mind the need for ease for dismantling unwanted or failed equipment and the need for ready identification and recycling of such component materials.

7.13 DESIGNING TO SELL

With the exception of a very limited range of products, almost certainly associated with military or space equipment, the products of the design and materials engineers deliberations must be sold and, as has been emphasized earlier, these products must achieve the required performance at a minimum cost. An informed dialogue between the design and materials engineers is essential and should cover not only the selection of the optimum materials, but also the processes to be used in the course of production.

To meet the challenge facing those responsible for design:

(i) designs should be finalized only after all the alternatives available concerning dimensions, materials and their properties, and production routes have been subjected to comprehensive evaluation in terms of present costs and future trends

(ii) The key is frequently simplicity for fewer components usually result in lower cost and weight, and in greater reliability

In the preceding chapters the technical aspects of metals and their use have been examined, the intention being to enable the design engineer to participate fully in the discussions, which the complexity of modern materials and associated processes that makes it essential he holds with the materials engineers, so that their partnership will result in designs that fulfil both the technical and economic objectives.

All too frequently, the design engineer relies on the manufacturing brochures and the properties quoted in those brochures for the selection of a material to be used for the manufacture of a component with no appreciation of how these properties originate or how the method of manufacture will affect those properties.

It is the hope of the authors of this book that a design engineer reading it will be encouraged to consult his materials engineer colleagues or associates at a very early stage in the design process and be able to enter into a meaningful discussion with them which will result in an economic functional design using materials to the best advantage.

Index